2015

ARDUINO MANUAL IN ENGLISH TOMO I

Who should read this book

Build cars and remote control helicopters, manufacture different types of intelligent robots, create synthesizer sounds, pitching a complete weather station (with sensors of temperature, humidity, pressure ...), assemble a 3D printer, monitor the effectiveness of our beer cooler from the garden, controlled via Internet commissioning of heating and lights of our house when we're away from it, periodically send data domestic water consumption to our Twitter account, designing clothing that lights up to the presence of gas, establish a system of shock sequence as a password to open doors automatically close off all televisions at once, implement an automatic irrigation system and self-regulated by state detected moisture in the soil, develop a theremin ray of light, making a musical alarm clock, using a video camera and radar for intrusion alarms on your mobile phone, play tic tac toe using spoken commands, etc. All this and much more can be achieved with Arduino.

This book is aimed, therefore, to anyone who wants to investigate how to connect the external physical world with the world of electronics and computers in order to achieve an autonomous and almost "intelligent" interaction between the two worlds. Engineers, artists, teachers and amateurs can learn about the possibilities offered by the Arduino ecosystem to perform almost any project that proposes imagination.

This course is designed for users with no knowledge of programming and electronics. It is assumed that the reader has a basic level of home computing (for example, knows how to unzip a "zip" or how to create a shortcut file) but no more. Therefore, this paper is ideal for anyone who has never programmed or has made any electrical circuit. In a way, thanks to the "excuse" of Arduino, what the reader has in his hands is a manual initiation both electronics as the basic programming.

The text is written self-taught reader facilitating a gradual assimilation of the concepts and procedures necessary to move forward gradually and safely throughout the different chapters, from the first to the last. This structure makes the text also can be used perfectly as a reference book for teachers who lecture Arduino in various areas (secondary education, vocational training, unregulated workshops, etc.). Spiced with many examples of circuits and codes, reading comprehension allows Arduino universe in a practical and progressive.

However, although very full, this course is not a reference or exhaustive compendium of all the functionality offered by the Arduino system. It would be impossible to cover them all in one volume. The experienced reader will note that in the following pages are missing to name and explain advanced aspects such interesting (some of which a whole book can lead themselves) as the Arduino role in building robots and 3D printers, or the possibilities of communication between Arduino and Android system devices, for example.

This course has been written taking into account several aspects. We have tried as far as possible to write a manual that is self-contained and progressive. That is, it is not necessary to resort to external sources of information to understand everything that is explained, but the text itself is self-explanatory in itself. And besides, all the information set out is displayed in an orderly and graduated without introducing concepts or procedures previously unexplained. Therefore, we recommend a sequential reading from the first chapter to the last, without jumps.

The methodology used in this text is largely based on exposure and detailed explanation of many examples of short code and concise: it has tried to avoid lengthy and complex codes, although interesting and colorful, can distract and disorient the reader to be too boundless. The idea is not to present complex projects already finished, but to expose the simplest possible form the basics. In this sense, many links are provided to extend the knowledge they do not have space in the book there are many issues (electricity, electronics, algorithms, mechanics, acoustics, electromagnetism, etc.) are proposed for the reader who has initiative to investigate on their own.

The chapter structure is as follows: the first section introduces the basic concepts of electricity in electronic circuits, and describes examples concretos- -through the performance and utility of the components present in most of these circuits (such as resistors , capacitors, transistors, prototyping plates, etc.). The second chapter presents the different plates forming the Arduino ecosystem components that are most important and concepts associated with this platform. The third chapter shows the programming environment official Arduino and describes its installation and configuration. The fourth chapter reviews the basic functionality of the Arduino programming language, offering numerous examples where you can see different flow structures, functions, data types, etc., used by this language. The fifth chapter shows the diversity of official libraries that incorporates the Arduino language and fail to deepen management hardware that uses them (SD, LCD screens, motors, etc.). The sixth chapter focuses on managing the inputs and outputs of the Arduino, both analogue and digital, and manipulation through buttons or knobs, among others. The seventh chapter explains various types of sensors via wiring examples and code; between treaties sensors are light sensors, infrared distance, motion, temperature, humidity, atmospheric pressure, and flexural strength, sound ... The eighth and last chapter analyzes the ability of Arduino boards to communicate with other devices (such as computers, mobile phones or other Arduino) using TCP / IP wired or wireless (Wi-Fi) networks, and Bluetooth, in addition to proposing a multitude of practical examples of interaction and data transmission through network.

Basic Electronics

THEORETICAL CONCEPTS ON ELECTRICITY

What is electricity?

An electron is a subatomic particle that has a negative electrical charge. Therefore, due to the physical law attraction together electrical charges of opposite sign (and repulsion of electric charges from each of same sign), any electron is always attracted to a positive charge equivalent.

A consequence of this is that if, for reasons that study, at one end (also called "pole") of a conductive material excess electrons appears at the other pole and a lack of these (equivalent to the existence of listed "positively charged"), electrons tend to move through that conductor from the negative to the positive. To this flow of electrons through a conductor material it is called "electricity."

Electricity exist as an offset load is not reached between the two poles of the driver. That is, as electrons move from one end to the other, the negative will become less negative and the positive pole will be increasingly positive, until the moment in which both sides have an overall neutral charge (say, are in equilibrium). At this situation, the movement of electrons cease. To avoid this, in practice they often use a external power supply (which is called a "generator") to constantly restore the initial difference in charges between the ends of
conductor, like a "bomb". Thus, while the function generator, the displacement of electrons can continue without interruption.

What is the voltage?

In the study of the phenomenon of electricity there is a fundamental concept that is the voltage between two points of an electrical circuit (also called "tension", "potential difference" or "potential drop"). Expliquémoslo with an example.

If two points of a conductor there is no difference of electric charges, the voltage between the two points is zero. If a charge imbalance (ie, at one point there is an excess of negative charges and the other an absence) appears between these two points, a voltage between the two points, which will be higher as it will appear that the difference load is also increased. This voltage is responsible for the generation of electron flow between the two points of the driver. However, if the two points have a charge imbalance together but are connected by a nonconductive material (what is called an "insulating" material), there will be a voltage between them but no passage of electrons (ie, not will electricity).

Generally, they say that the point of the circuit with large excess of positive charges (or put another way: more lack of negative charges) is the one with the highest "potential", and point more excess negative charges it is the one with the smallest "potential". But never forget that the voltage is always measured between two points does not make sense to say "the voltage at this point," but "the voltage at this point on this one"; hence its other name "potential difference" or "potential drop".

So, as we will use will always be potential differences between two points on the absolute numerical value of each of them we can allocate as appropriate. That is, although 5, 15 and 25 are different absolute values, the potential difference between one point and another worth 25 worth 15, and the difference between one and another worth 15 worth 5 gives the same result. For this reason, and for convenience and ease of calculation, the point of the circuit with smaller potential (the highest negative charge, remember) is usually given a reference value equal to 0.

Also by agreement (although physically is actually just the opposite) they say that the electrical current going from the point with the greatest potential to another point with lower potential (that is, the accumulated charge at the positive end is moving towards the negative end).

To better understand the concept of voltage can use the analogy of the height of a building: if we assume that the point with the smallest potential is the ground and assume this as the reference point with value 0 as an elevator go up the building will acquire more and more about the ground potential: the more you have the lift height, will be more potential difference between it and the ground. When we are talking about a "potential drop", then we will want to say (in our example) that the elevator has reduced its height above the ground and therefore has a lower voltage.

The unit of measurement of voltage is the volt (V), but can also speak millivolt (1 mV = 0.001 V) or kV (1 kV = 1000 V). Typical values for home electronics projects as we will discuss in this book are 1.5 V, 3.3 V, 5 V ... although when involved mechanical elements (such as engines) or other complex elements will need to bring something more power to the circuit, so the values are somewhat higher: 9 V, 12 V or 24 V. In any case, it is important to note that values beyond 40 V can endanger our lives if we do not appropriate precautions; projects in this book, however, no voltages of this magnitude is never used.

What is the current intensity?

The current intensity (commonly called "current" for short) is an electrical quantity which is defined as the amount of electric charge passing in a given time through a particular point of a conductive material. We can imagine that the current is similar in some ways to the flow of water flowing through a pipe: to spend more or less amount of water through the pipe at a certain time would be analogous to spend more or less amount of electrons by a electric cable at the same time.

Its unit of measure is the ampere (A), but can also speak milliamp (1 mA = 0.001 A) of microamperes (uA 1 mA = 0.001), or even nanoamp (1 NO = 0.001 uA).

As we mentioned, it is generally considered that a circuit current flows from the positive terminal (point of highest tension) to the negative (point lower voltage) through a conductive material.

What is direct current (DC) and alternating current (AC)?

Two basic types of circuits must distinguish when we speak of magnitudes such as voltage or current: DC circuit (or DC circuits, the English "Direct Current") and AC circuits (or AC circuits, the English " Alternating Current ").

Current call to one in which the electrons flow through the conductor always in the same direction (ie, in which the ends of major and minor, or potential that is, the positive and negative-pole are always the same). Although commonly direct current with constant current is identified (for example, supplied by a battery), strictly alone is continuous any current that, as we have said, always keep the same polarity.

We call alternating current one in which the magnitude and polarity of the voltage (and therefore also the intensity) vary cyclically. This implies that the positive and negative poles are alternately exchanged over time and therefore, the voltage is taking positive and negative values with a certain frequency.

Alternating current is the type of power that reaches homes and businesses from the electricity grid. This is because the AC is easier and more efficient to transport over long distances (as it suffers less energy loss) than continuous current. In addition, the alternating current can be converted to different voltage values (either increasing them or decreasing them as we are interested through a device called a transformer) more easily and efficiently.

However, all the projects in this book use only current as the circuits where we can use Arduino (and indeed, most home electronics) only work correctly with this type of power.

What is the electrical resistance?

We can define the internal electrical resistance of any object (although usually we refer to any electronic component that is part of our circuits) and its ability to oppose the passage of electric current through it. That is, the higher the resistance of that component, electrons will have more difficulty to cross it until the end even preclude the existence of electricity.

This feature depends among other factors of the material from which the object is built, so we can find materials with little or very little intrinsic resistance (called "drivers" such as copper or silver) and materials with enough or too much resistance (called "insulators" like certain types of wood or plastic, etc.). However, it should be stressed that although a conductive material is always inevitably possess an inherent resistance that prevents 100% of the power is transferred through it, so even a simple copper wire has some internal resistance (usually negligible, of course) that reduces the flow of the original electrons.

The unit of measurement of the resistance of an object is the ohm (Ω). We can also talk about kilohms (1 kW = 1000 Ω) of megohms (1 M = 1000 KQ), etc.

What is Ohm's Law?

Ohm's Law says that if an electrical component with internal resistance, R, is crossed by a current, I, between the two ends of said component be a difference of potential, V, which can be known by the relationship V = GO.

This formula is easy to deduce interesting relationship of proportionality between these three electrical quantities. For example it can be seen that (assuming that the internal resistance of the component does not change) the greater the intensity of current passing through it, the greater the potential difference between its ends. It can also be seen that (assuming in this case that at all times there flows the current intensity by the component), the higher internal resistance, the higher the potential difference between its two ends.

In addition, clearing the appropriate quantity of the above formula, we can obtain, from two known any data, the third. For example, if we know V and R, we can find I using I = V / R, and V and I know if we can find R by R = V / I.

From the above formulas should be also easy to see for example that the higher the voltage applied between the ends of a component (which we assume that has a fixed resistance value), the greater the intensity of

current through it. Or that the greater the strength of the component (keeping constant the potential difference between its ends), the lower the intensity of current passing through it. In fact, in the latter case, if the resistance value is high enough, we can get even flow of electrons is interrupted.

What is power?

We can define the power of an electric / electronic component as the energy consumed by this in a second. If, however, we are talking about a power source with the word power we then refer to the electrical energy supplied by this circuit in a second. In both cases (either consumed or generated power), power is an intrinsic value of the component or generator, respectively. Its unit of measurement is the watt (W), but can also speak of milliwatt (1 mW = 0.001 W) or kilowatts (1 kW = 1000 W).

From the very known power component / generator and the time that this is working, you can know the total consumption / energy provided by the expression: $E = P \cdot t$.

When a power source provides a given power, it can be consumed by various components of the circuit in various ways: in most cases is wasted as heat due to the effect of the inherent internal resistance of each component (called " Joule effect "), but can also be consumed as light (whether that component is a lamp, for example) or in the form of motion (if that component is a motor, for example) or as sound (if that component is a speaker, for example), or a mixture of several.

We can calculate the power consumed by an electrical component if we know the voltage to which it is subject and the intensity of current through it, using the formula $P = V \cdot I$. For example, a light bulb under 220 V for circulating 1A consume 220 W. On the other hand, from Ohm's Law can deduce two equivalent formulas can be useful if we know the internal resistance value R Component: $P = I2 \cdot R$ or also $P = V2 / R$.

Finally, you should know that conductive materials can support up to a maximum amount of power consumed, beyond which there is a risk of overheat them and damage them.

What are digital signals and analog signals?

We can classify the electrical signals (either voltages or currents) in various ways according to their physical characteristics. One possible classification is to distinguish between analog signals and digital signals.

Digital signal is one that only has a finite number of possible values (what is often called "having discrete values"). For example, if we consider the color signal emitted by a traffic light, it is easy to see that this is digital, because it can only have three specific, distinct and smooth transition without the possibility of including values: red, amber and green.

A particular case of digital signal is the binary signal, where the number of possible values is only 2. Know this type of signal is important because electronics is very common to work with voltages (or currents) with only two values. In these cases, one of the binary voltage values usually 0 Or a aproximado- value to indicate precisely the absence of voltage, and the other value can be anything, but sufficiently distinguishable from 0 to indicate unambiguously the presence of signal. Thus, a value of binary voltage always identifies the state "no current" (also called the "off"
- "Off" in English, UNDER -LOW in English, or "0") and other value provided identifies the status "current flows" (also called "on" - "on" - HIGH -HIGH - or "1").

The concrete voltage value corresponding to the HIGH state will be different according to the electronic devices used at all times. In projects of this book, for example, will use common values of 3.3 V or 5 V But beware: it is important to note that if we submit an electronic device at too high a voltage (for example, if we apply 5V as high when the device supports only 3.3 V) we risk irreversibly damaging.

In addition to high and low levels, there is a binary signal transitions between these levels (from HIGH to LOW and LOW to HIGH), called flank of downstream and upstream, respectively.

Analog signal is one that has infinite possible values within a certain range (which is often called "having continuous values").
physical quantities (temperature, sound, light, ...) are analog, as well as more specifically the power (voltage, current, power ...) because all of them, naturally, can undergo continuous variation without jumps.

However, many electronic systems (a computer, for example) do not have the ability to work with analog signals: can only handle digital signals (especially binary type, hence its importance). Therefore, they need an analog-digital converter that "translate" (or rather, "simulating") the analog signals from the outside world into meaningful digital signals by said electronic system. A digital-analog converter is also required if you want to perform the inverse process: transform a computer's internal digital signal into an analog signal so it can be broadcast to the physical world. An example of the former would be recording a sound using a microphone, and one of the latter would be playing a pre-recorded sound through a speaker.

On the methods used to perform these conversions from analog to digital, and vice versa, we'll talk at length later, but what we know is because, whatever the method used, there will always be a loss of information ("quality") during the process of signal conversion. This loss occurs because it is mathematically impossible to make a perfect transformation of an infinite number of values (analog signal) to a finite number (digital signal) because, by force, several values of the analog signal should "collapse" into a single value indistinguishable from the digital signal.

Despite the above, the reason why most electronic systems used to operate digital signals instead of analog is because the former have a great advantage over the second: they are more immune to noise. By "noise" means any unwanted signal variation and is a phenomenon that occurs constantly due to a multitude of factors. The noise modifies the information that provides a signal and greatly affects the correct functioning and performance of electronic devices. If the signal is analog, the sound is much more difficult to treat and recover the original information is complicated.

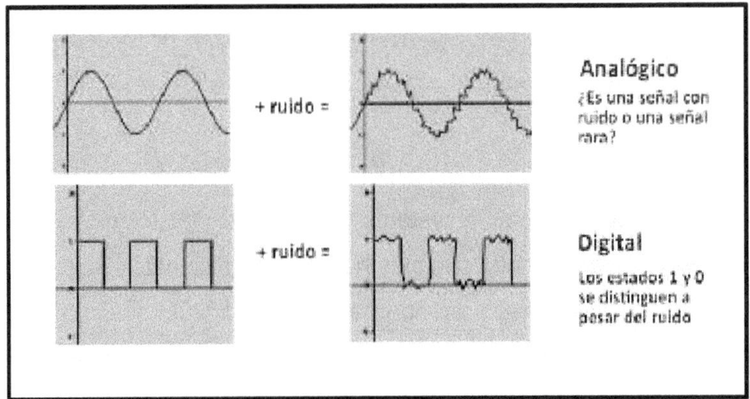

What are periodic signals and aperiodic signals?

Another classification we can do with the electrical signals is split between periodic and aperiodic signals. We call that periodically repeated after a certain period of time (T) and aperiodic signal that non-repeating signal. In the case of the first (the most interesting with difference), depending on how the signal varies over time, this can have a particular "shape" (-es sine say, following the drawing seno- function, square, triangular, etc.).

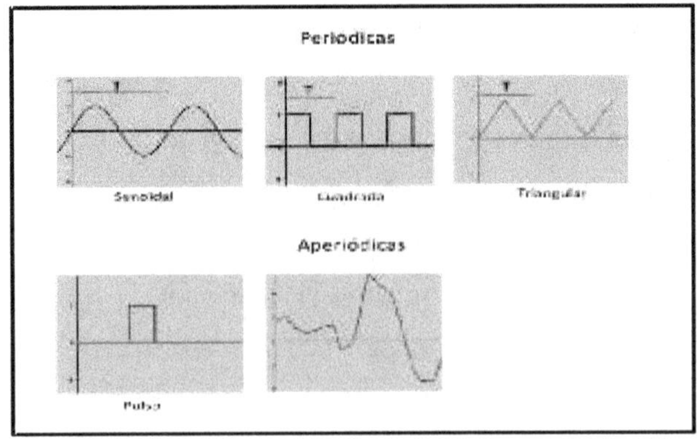

The periodic signals have a number of characteristics that we identify and set to work with them in a simple way:

Frequency (f) is the number of times the signal repeats in one second. It is measured in hertz (Hz), or multiples thereof (such kilohertz or megahertz). For example, if we say that a signal is ten hertz, meaning that repeated ten times each second.

Period (T) is the duration of a full cycle of the signal before it repeats again. It is the inverse of the frequency (T = 1 / f) and is measured in seconds.

Instantaneous value: the specific value that takes the signal (voltage, current, etc.) at each instant

Average value: it is a mathematically calculated value by averaging the different values that the signal has been given over a specific time. Some electronic components (for example, some engines) does not respond to the instantaneous value but the average value of the signal.

BASIC ELECTRICAL CIRCUITS circuit Graphing

To describe simply and clearly the structure and composition an electric circuit diagrams are used graphics. They each circuit device is represented by a standardized and all interconnections between them are drawn symbol. For example, a simple circuit would be:

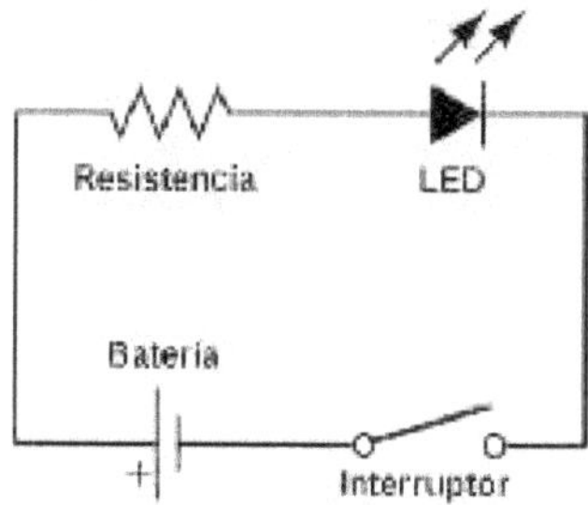

In the above scheme we see four devices (present in virtually any circuit) represented by the conventional symbol: A

cell or battery (whose task is supplying power to the other components), a resistor (component specifically designed to resist the passage of current, hence its name), an LED (component that lights up when receiving power) and a switch. In this example, the battery will create the potential difference necessary between its two ends, also called "terminals" or "poles" - to which an electrical current is generated, which emerge from the positive terminal (marked with the sign "+ "), it will pass through the resistance, then pass through the LED (illuminating, therefore) and will reach its final destination (the negative pole of the battery) as long as the switch closes the circuit.

Clarify what "closed circuit". We just say that an electric current always flows from the positive pole of the battery to the negative if there is a potential difference appears. But this is only possible if there is a path between the two poles (the circuit itself) to permit the passage of said stream. If the circuit is open, even though the battery is working, current will not flow. The function of the switches is precisely close or open the circuit so that current can pass or not, respectively. In the following diagram this is clearer:

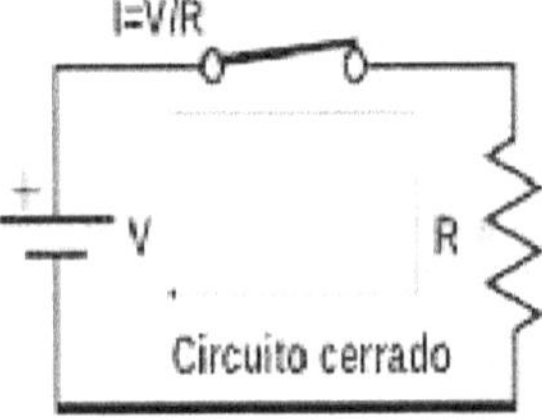

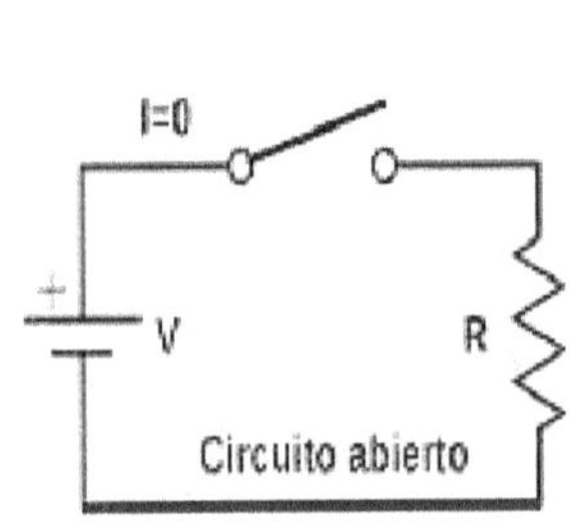

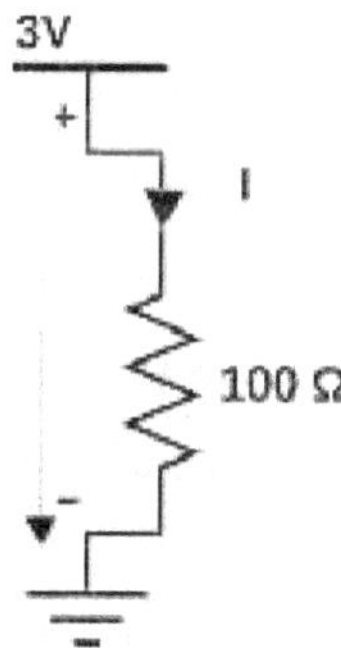

On the other hand, the circuits can be represented alternatively in a slightly different way to that shown above, using the concept of "land" (also called "mass"). The "earth" ("ground") is just one point of the circuit arbitrarily chose as a reference for measuring the potential difference between this and any other point of the circuit. In other words, the point where we say that the voltage is 0. For practical use, the point of land normally associated to the negative pole of the battery. This new concept we often simplify the drawing of our circuits as

if we represent the ground point with the ⏚ symbol, the circuits can draw

as follow:

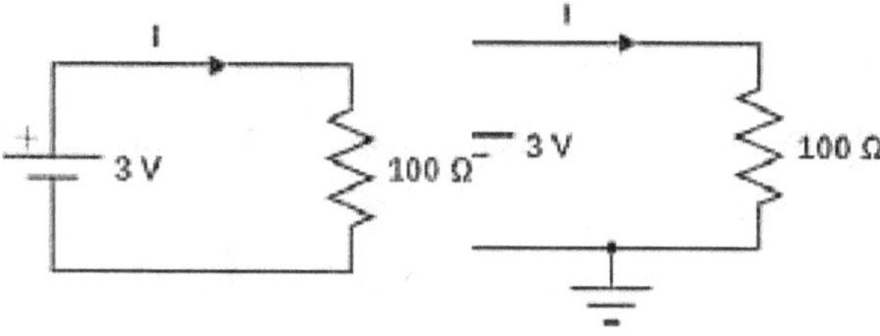

We may also meet with wiring diagrams that show intersections cables. In this case, we look if it appears drawn a circle at the center point of the intersection. If so, we are indicating that the cables are physically and electrically connected together. If it does not draw any circle at the center point of the intersection, we are indicating that the cables are simply independent pathways that intersect in space.

Serial and parallel

The various devices in a circuit can be connected together in various ways. The most basic are the "serial connection" and "Parallel Connection". In fact, any other type of connection, however complex, is a combination of any of these two.

If various components are connected together in parallel, they are all the same voltage applied equally (ie, each component works the same voltage). Moreover, the total current is the sum of the currents passing through each component, as there are several possible paths for the passage of electrons.

If connected in series, the total voltage will be distributed (usually unevenly) between different components, so that each work under a part of the total voltage. That is, the total voltage is the sum of the stresses in each component. On the other hand, the intensity of current flow through all components in series will always be the same, since there is only one possible way for the passage of electrons.
You can better understand the difference by the following schemes, in which you can see the connection in series and in parallel of two resistors.

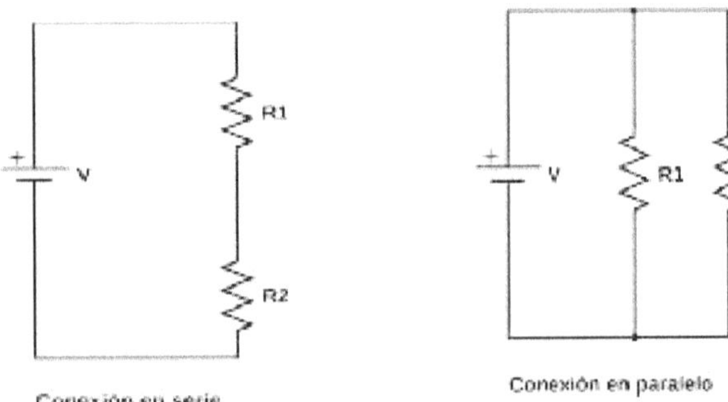

Conexión en serie

Conexión en paralelo

With Ohm's Law we can get the value of any electrical quantity (V, I or R) if previously know the value of any other involved in the same circuit. To do this, we must take into account the particularities of the connections in series or in parallel.

Consider this example using the circuit of the two resistors in series:

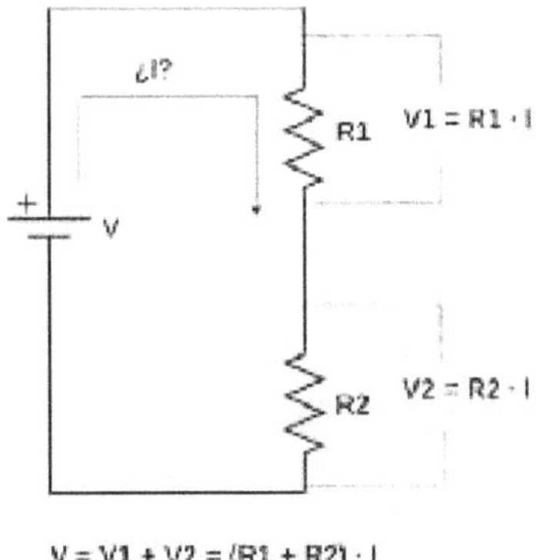

In the above scheme represents the applied voltage V1 to the voltage V2 R1 and R2 applied. If we have for example, a power supply (battery) which provides a voltage of 10 V and two resistors whose values are R1 = R2 = 1Ω and 4Ω respectively, to calculate the current flowing so as R1 R2 (recall that it is the same because there is only one possible way) simply should perform the following operation: I = 10 V / (1 + 4 Ω Ω) = 2 a, as shown in the above scheme.

Consider now the circuit of the two resistors in parallel:

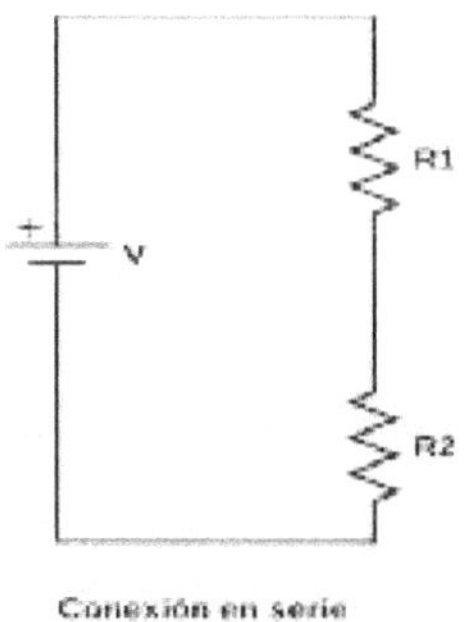

Conexión en serie

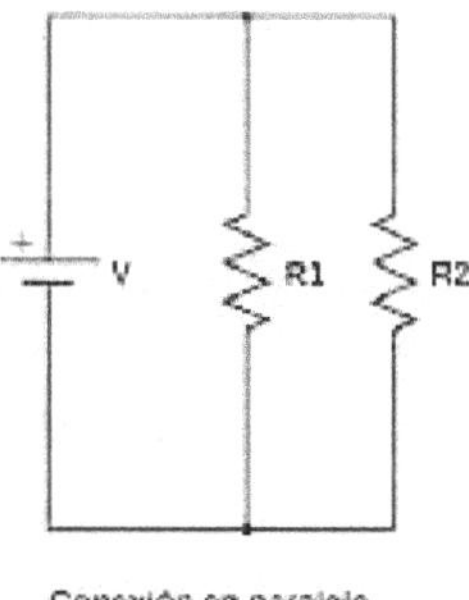

Conexión en paralelo

With Ohm's Law we can get the value of any electrical quantity (V, I or R) if previously know the value of any other involved in the same circuit. To do this, we must take into account the particularities of the connections in series or in parallel.

Consider this example using the circuit of the two resistors in series:

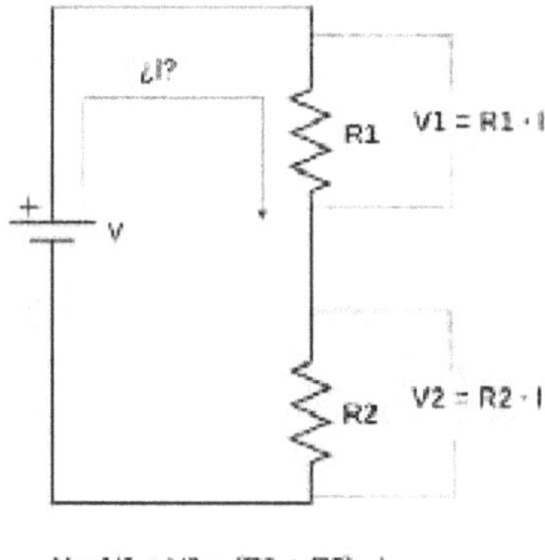

V = V1 + V2 = (R1 + R2) · I

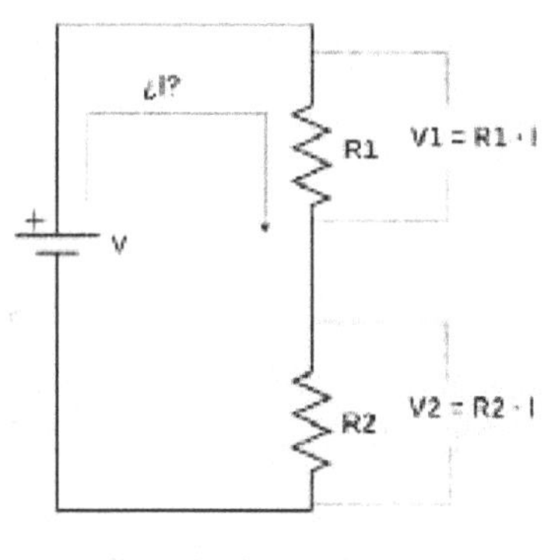

V = V1 + V2 = (R1 + R2) · I

In the above scheme I1 represents the intensity of current through R1 and I2 the intensity of current through R2. If we have for example, a power supply (battery) which provides a voltage of 10 V and two resistors whose values are R1 = R2 = 1Ω and 4Ω , respectively, to calculate the current flowing through R1 should perform (as shown in the diagram) the following operation: I1 = 10 V / 1 Ω = 10A; to calculate the current through R2 should do: I2 = 10 V / 4 Ω = 2.5A; and the total current flowing through the circuit is the sum of the two: I = I1 + I2 = 10 A = 12.5 + 2.5 A A.

From the above examples, we can deduce a pair of formulas come us well along the whole book to simplify the circuits. If we have two resistors connected in series or in parallel, it is possible to replace them in our calculations by a single resistor whose behavior is completely equivalent. For the serial connection, the value of said resistance (R) would be given by $R = R1 + R2$, and in the case of parallel connection, the equivalent value is calculated using the formula $R = (R1 \cdot R2) / (R1 + R2)$, as can be seen in the following diagram.

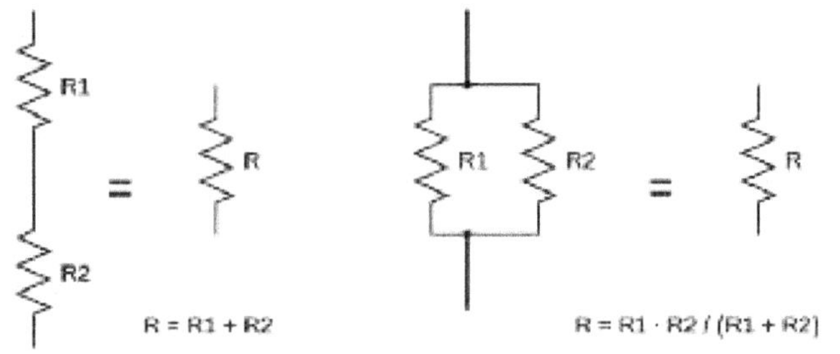

An interesting fact to take into account (which is deducted from the formula itself) is that when resistors are connected in parallel, the resulting value of R is always less than the lowest value of the resistance involved.

The voltage divider

The "voltage divider" is nothing more than a circuit formed by a resistor connected in series with any other electrical device. His intention is to reduce the voltage applied to the device, setting it to a safe level to avoid damaging it. In other words: the "voltage divider" is to obtain a lower voltage that some original voltage.

The greater or lesser amount of reduction we get into the final voltage depends on the resistance value that we use as a divider: the higher the resistance value, the greater reduction. However, keep in mind that the voltage obtained also depends on the value of the original tension: if we increase this, that also will increase proportionally. All these values can be calculated using a concrete example, as the following scheme.

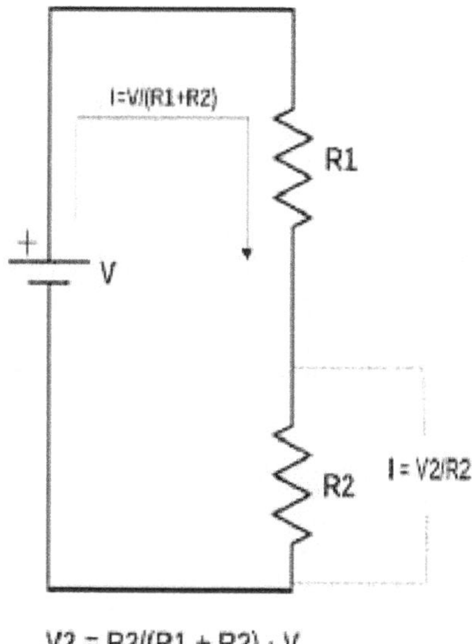

As can be seen, we have a power supply (battery) which provides a voltage of 10 V and two resistors whose values are R1 = 1Ω (which will voltage divider) and R2 = 4 Ω, respectively. We also know that the current I is always the same at all points of the circuit-and no ramifications parallelogram. Therefore, to calculate V2 (ie, the voltage applied to R2, which has been lowered from contributed by the battery through R1), we can realize that I = V2 / R2 and I = V / (R1 + R2), so here it is easy to obtain that V2 = (R2 · V) / (R1 + R2). It is therefore clear expression

above what was said in the previous paragraph: V2 always be proportionally less than V, and as more R1, V2 will be lower.

Resistance "pull-up" and "pull-down"

Many times, the electrical circuits are "tickets" by receiving an electrical signal from outside (binary type) that has nothing to do with the signal obtained from the power source. These external signals can be used to many things: to enable or disable parts of the circuit, the circuit to send information from their environment, etc.

Resistance "pull-up" (and "pull-down") are normal resistance, just bear that name for the role: they serve to assume a default value of the signal received at an input of the circuit when it is not No specific value (neither tall nor short), which is what happens when the input is not connected to anything (ie, is "on the air") is detected. Thus, this type of resistance ensure that the binary values fluctuate received no nonsense in the absence of input signal.

In resistance "pull-up" the value is assumed by default when no external device signal emitter connected to the input is high and the "pull-down" is low, but both have the same objective, so the choice of a resistance-type "pull-up" or "pull-down" depend on the particular circumstances of our assembly. The difference between each other is in their location within the circuit: resistance "pull-up" are connected directly to the external signal and the "pull-down" directly to earth (see diagrams below).

Consider a concrete example of the usefulness of a "pull-down" resistance. Suppose we have a circuit as follows (where the 100 ohm resistor is merely placed a voltage divider circuit at the input to protect).

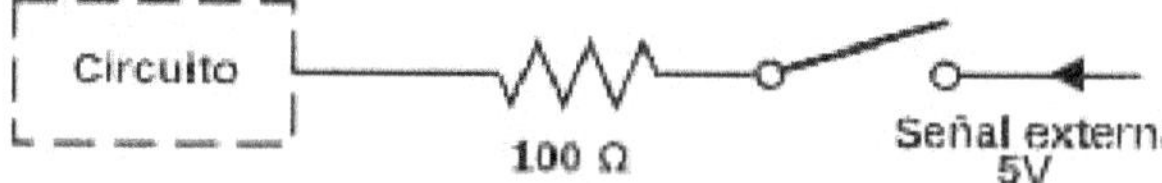

When the switch is pressed, the input circuit is connected to a valid input signal, which assume binary (ie, you have two possible values: HIGH-of 5V, for example-and LOW-of 0V-), so the circuit will receive one of these two specific values and everything will be ok. However, if the

interruptor se deja de pulsar, el circuito se abrirá y la entrada del circuito no estará conectada a nada. Esto implica que habrá una señal de entrada fluctuante (también llamada " flotante" o " inestable") que no nos interesa. La solución en este caso sería colocar una resistencia " pull-down" así:

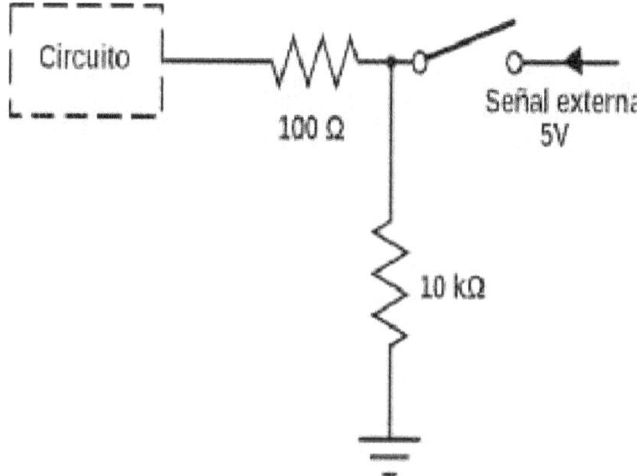

In this example the "pull-down" resistance is 10 k. When the switch is pressed, the input circuit is connected to a valid input signal, as before. When the stop switch is pressed, the input circuit is connected to the "pull-down" resistor, which pulls earth (which is a reference always fixed).

One might think that when the switch is pressed, the circuit will receive the input signal but also be grounded through the resistance "pull-down": what really happens then? Here's the key to why the "pull-down" resistor is used and no direct connection to earth is used: the opposition to the passage of electrons from the external signal that carries the "pull-down" resistance causes these are always diverted to the circuit input. If we input connected to ground directly without using the "pull-down" resistor, the external signal would head directly to ground without passing through the input of the circuit because that way would find least resistance (Ohm pure Act: less resistance plus intensity).

With resistance "pull-up" he could have gotten the same, as shown in the following diagram. In this case, when the switch is pressed outward sign that land is diverted to find a direct path to it (so the input circuit receives nothing-a "0" -) and when the switch is left without pressing when the input circuit receives the external signal. Be careful with this.

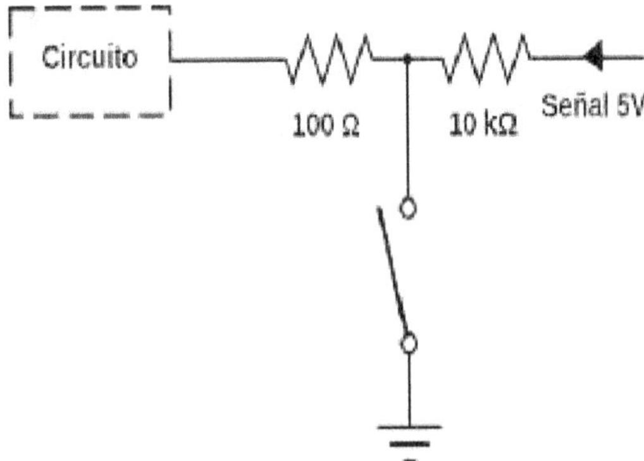

In the above examples we used resistance "pull-up" or "pull-down" of 10 k. Is a fairly common standard used this particular value electronics projects where work is in the range of 5V, but, in any case, if we tune it somewhat, we can calculate the ideal value using Ohm's Law from the current consuming circuit.

SOURCES OF POWER SUPPLY Types of batteries / battery

We call power source to charge element
generating the potential difference necessary for the flow of electric current through a circuit and thus can operate the devices connected to it. The sources we use most often in our projects are of two types: the cells or batteries and AC / DC adapters.

The term "battery" is used to denote generators typically not based on reversible chemical processes electricity and therefore are not rechargeable generators; while the term "battery" is usually applied to semi-reversible electrochemical devices that allow be recharged, although these terms are not strict formal definition. The term "battery" applies equally to one kind or another (as well as other types of voltage generators, and electrical capacitors) as being a neutral term that encompasses and describe them all.

If we distinguish the battery / batteries internal chemical dissolution responsible for the generation of the potential difference between the poles, We find that the most widespread in the market currently batteries ("non-rechargeable batteries") are the alkali type, and the most common batteries ("rechargeable batteries") are on the one hand the nickel-cadmium (Ni-Cd) and all nickel-metal hydride (NiMH), and secondly the lithium-ion (Li-ion) and ion polymer battery (LiPo). Of all these types of batteries, Lipo are those having a higher charge density (ie, that being the lightest are those, however, more autonomy) but are more expensive.

The industry is international standardization common standards for the manufacture of alkaline batteries and batteries Ni-Cd / NiMH which define certain sizes, shapes and predetermined voltages, so that they can be used without problems in any electrical appliance worldwide. In this regard, the most common types of batteries are type D (LR20), C (LR14), AA (LR06) and AAA (LR03), each 1.5 V and generating cylindrical but of different dimensions (in fact, they have larger list to smallest). There are also frequent PP3 (6LR61) type, which generate 9 V and have a rectangular prism; 3R12 and of type ("pouch") that generate 4.5 V and have flattened cylindrical shape. The following image shows, from left to right, -alcalinos- accumulators type D, C, AA, AAA, AAAA and PP3, placed on graph paper.

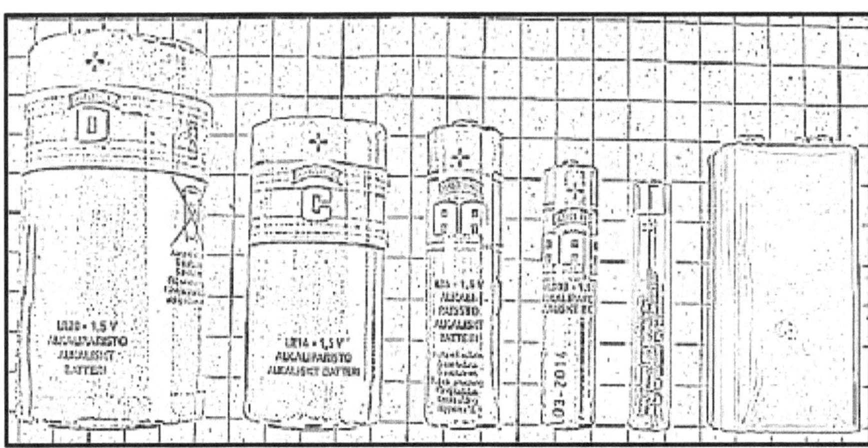

In the picture below, left two batteries of LiPo type and right two encapsulated made of cylindrical lithium-ion type display. The former usually come in the form of thin rectangles in a silver bag and the latter usually come in a rectangular or cylindrical hard case, although both actually come in a wide versatility and flexibility of shapes and sizes. The LiPo are lighter than the Li-ion but often have a lower capacity, so the first is usually used in small devices such as mobile phones and the second in portable chargers and the like.

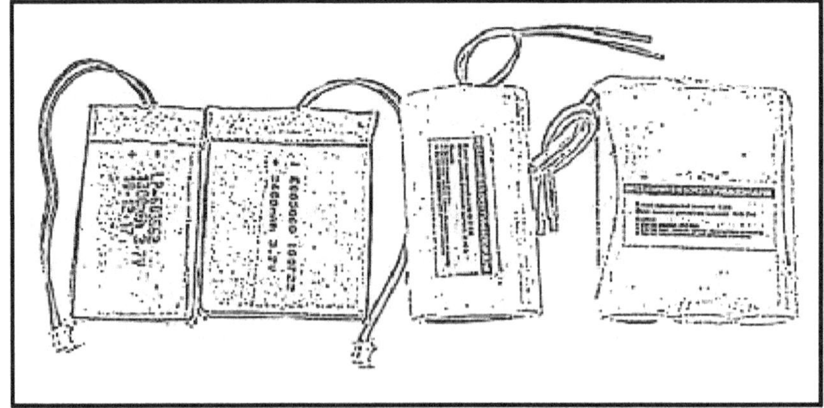

We must also indicate the existence of such batteries "button". There are many types: if they are made from lithium-manganese dioxide, its nomenclature starts with "CR" (well, we have the CR2032, the CR2477, etc.) and while each has an encapsulated diameter and width different, all generate 3V If they are made of silver oxide nomenclature commonly starts with "SR" or "SG" (well, we have the SR44, the SR58, etc., depending on its size). There are also alkaline type, whose code commonly starts with "LR" or "AG". Both silver oxide as alkaline generate 1.5 V. In any case, whatever kind it is, in all the negative battery terminal button is the cover and the positive terminal is the metal of the other side.

Features battery / batteries

Keep in mind that the voltage is different batteries provide a "nominal" value: that is, for example a 1.5 V AA battery really early in his life generates about 1.6 V, quickly drops 1.5 V and then gradually goes down to 1 V, at which the battery can be considered the "tired". The same goes for other types of battery; for example, a battery Lipo marked "3.7 V / (4.2 V)" indicates that initially is able to provide a maximum voltage of 4.2 V but quickly drops to 3.7 V, which is the voltage means for most of his life, and finally down to 3 V quickly and automatically stop working. In this regard, it is useful to consult the official documentation provided by the manufacturer for each particular battery (called "datasheet" battery) to see voltage variation provided depending on the operating time.

In addition to the generated voltage of a battery / battery (which will assume from now always constant) we must know another important feature: the electric charge that this is capable of storing (sometimes called "capacity" battery / battery). This value is measured in ampere-hours (Ah) or milliampere-hour (mAh) and lets us know approximately how much amperage can bring the battery / battery for a specific period of time. In this sense, we must remember that while the voltage supplied by the battery / battery is ideally constant, the intensity provided, however, varies from time to time as do the power consumption of the circuit to which we connect. For example, 1 means that in theory Ah battery / battery can deliver for an hour a current of 1 A (if required by the circuit) or 0.1 A for 10 hours, or 0.01 A for 100 hours, etc., but always at the same voltage.

However, the above is not exactly true, because the more amount of current the battery / battery supply, actually its operating time will be reduced to a much more marked by their ability proportion. For example, a button battery 1 Ah is unable to provide 1 A for an hour (or even 0.1 A at 10 hours) because it runs much earlier, but instead, has no problems to contribute 0.001 for 1000 hours. To find specific current intensity that enforces the nominal value of a battery Ah, we must consult the manufacturer's documentation (the "datasheet" battery). This intensity "optimal" current in the case of LiPo batteries is often called "Capability", and is expressed in units C, where C unit corresponds to the value of the battery Ah divided by one hour.

For example, the C drive of a battery with 2 Ah load is 2 A, and its concrete Capability will be a certain amount of C units, available on datasheet (1C, 2C ...). If we then, for example, a battery of 2 Ah and 0.5 C and another 2 Ah and 2C, the first will provide a steady stream of up to 1A without exhausting prematurely, and the second may provide a current of up to 4 A. known this, we must take into account for example that button batteries have a very small Capability (0,01C is a typical value), so if they are forced to bring a lot of intensity at a given moment, your life will be reduced drastically.

On the other hand, we have said that a battery supply at any given time a current or another basically depends on power consumption (measured in amperes, or more usually in milliamps) to conduct the full set of devices that are connected then the circuit (which can logically be very varied as appropriate). That is, the operating time of a battery depends on the demand of the circuit to which it is

connected. More specifically, we can get (very roughly) the download time of a battery / battery using the expression discharge time = battery capacity / power consumption circuit.

For example, if a battery has an electric charge of 1000 mAh and a device consumes 20 mA, the battery will take 50 hours of discharge; if instead the device consumes 100 mA, the battery will take only 10 hours to download. All this is the theory, since the numerical value of mAh printed on the battery should be taken only as an approximation, and should be considered only in ranges of consumption levels (measured in units C) specified by the manufacturer as high consumption and to know that this value can not be accurately extrapolated.

Connections of several battery / batteries

We have seen in previous diagrams symbol often used in the design of electronic circuits to represent a cell or battery is:

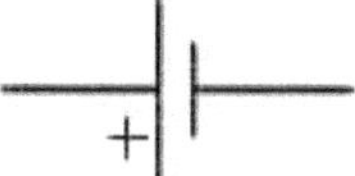

where the largest part (and sometimes painted thicker) the drawing represents the positive pole of the source. Often the "+" symbol is omitted.

When we talk about batteries connected "in series" we mean that we connect the negative pole of a positive pole of another, and so, so that we finally have an overall positive pole on one side and an overall negative on the other. In this figure you can understand better:

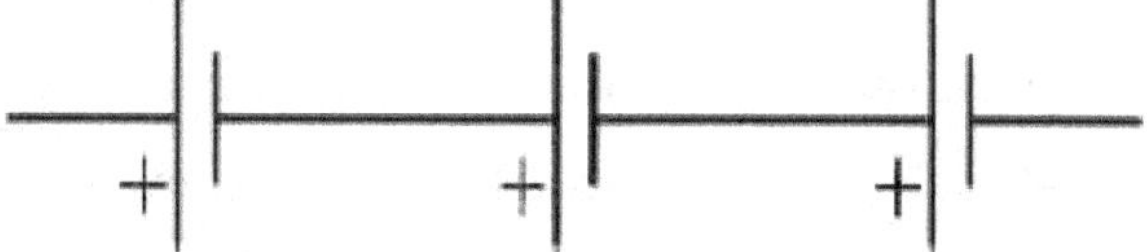

The series connection of batteries is useful when we need to have a battery that generates a certain relatively high voltage (e.g. 12 V) and single cells have lower voltage (eg, 1.5 V). In that case, we can connect these units of 1.5 V in series to obtain "by hand" the battery to provide the desired voltage, since the total voltage supplied by batteries connected in series is the sum of the individual voltages. In our example, for a 12 V from 1.5 V batteries, need 8 units, for 1.5 V · 8 = 12 V. In fact, commercial batteries 4.5 V and 9 V (and 6 V and 12 V, which is also the any) are usually manufactured in series internally connecting batteries 1.5 V. So many times we will see the following symbol (instead of the previously shown) representing a stack:

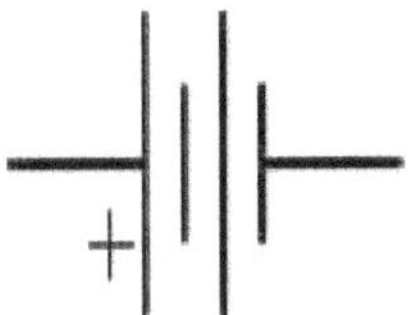

However, we must also take into account that the total capacity (ie, the assembly mAh batteries in series) does not increase: will remain exactly the same as having a battery that set of individually and independently. This is important because the circuits that need to be fed with high voltage generally have a higher power consumption, which (under the formula above) the running time of a source formed by cells in series is quite low.

Another way to interconnect different individual cells is in parallel: in this configuration, all poles of the same sign are linked together. That is, on the one hand the negative poles of each battery are connected and on the other hand all the negative terminals are connected, these two common connection points overall positive and negative poles.

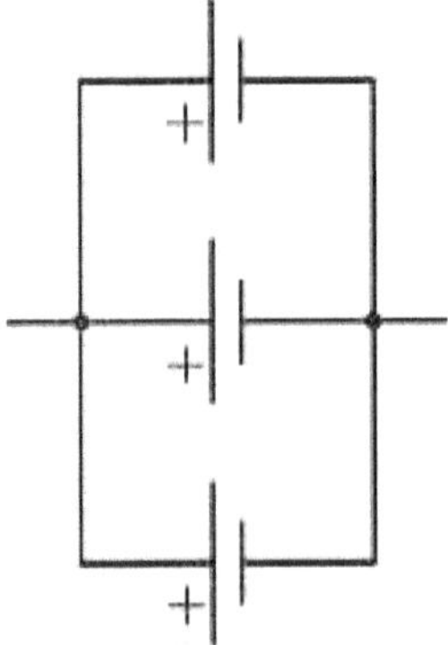

A parallel set of batteries provides the same voltage as one single stack (that is, if we for example four 1.5 V batteries connecte achieve is that the duration of maintaining the system voltage is greater than if we use a single stack, because the capacity (the mAh) of the set is the sum total of the capacities of each of the individual stacks.

It is very important to ensure that the battery / batteries connected in series or in parallel are the same type (alkaline, NiMH, etc.), are of the same form (AA, PP3, etc.) and provide the same voltage. If not done, the whole operation may be unstable and even dangerous: in the case of LiPo batteries can reach explode if this rule is not followed. In fact, this type of battery we recommend purchasing packs (serial or parallel) preassembled, as we offer a guarantee that their units have been selected to have the same capacity, internal resistance, etc., and not cause problems .d in parallel, this set also give a total voltage of 1.5 V). The advantage

Purchase batteries / battery

Any type of cell / battery we need in our projects (different voltage, capacity, chemical composition ...) it can be purchased through any of the dealers listed in Appendix A of the book, just use the search offering in their online stores to find the type of battery / battery required. For example, if we LiPo batteries, electronic components distributor Sparkfun provides, inter alia, Product No. 341 (a 3.7
V / 850 mAh) or No. 339 (3.7V / 1Ah), both with 2-pin JST connector. If we look for alkaline or NiMH rechargeable batteries, is even simpler: the most common models are available in any local store.

If indeed you are going to use this kind of (alkaline or rechargeable NiMH) batteries, the most common is to use in our projects some sort of holder that allows using multiple units connected in series. There are many models; as shown, only on the website of the electronic component distributor Adafruit can find (writing your product number in the integrated portal search engine), the following products:

No. 248: 6 AA battery holder units plug 5.5 / 2.1 mm.

875: "8 AA units plug of 5.5 / 2.1 mm and switch.
No. 727: "3 AAA units with connector JST 2 pin and switch.
No. 67: "1 PP3 unit with plug 5.5 / 2.1mm and switch.
No. 80: clip to attach a PP3 1 unit with plug 5.5 / 2.1 mm
No. 449: 8 AA battery holder units with direct cable (without connector). If desired, cables can be easily connected to a pin of 2.1 mm acquired separately (product # 369).
The 2.1 mm jack can come to us well to feed our Arduino as this has a female connector that. Furthermore, the product number 727 is designed to feed circuits AAA batteries originally designed to be powered by batteries Lipo, as these typically incorporate precisely JST type connector 2 pin.

Purchase Chargers

Another supplement that can come to us it is a battery charger. For example, for NiMH us it is sufficient that offered for example by SparkFun with product code No. 10052 This charger plugs directly into a wall outlet and can be used with AA, AAA or PP3 batteries.

But if we want to recharge LiPo batteries, not just any magazine, you have to be careful to always use one that meets two conditions: to provide a voltage (preferable) or less than the maximum voltage provided by that particular battery, and also that it makes a battery capability of monitoring the specific intensity equal to or less (preferable).

If the above conditions are not met, the charger may damage the battery irreversibly (and even to explode). LiPo batteries are very sensitive: it also runs the risk of explosion when discharged below a minimum voltage (typically 3 V), or when they are forced to provide more current than can be offered (usually 2C), or when they are used in environments with extreme temperatures (usually outside the range 0 ° C-50 °), among other causes. So many of these batteries (though not all) include a protective circuit that detects these situations and disconnects the battery safely. However, for specific characteristics of each battery (such as voltages, currents and safe temperatures) it is required to consult the information provided by the manufacturer for that particular battery (their datasheet).

The LiPo (and Li-ion) battery chargers may be found in many places and in many forms. As examples, only on the website of Sparkfun we can find the product number 10217, 10401, 11231 or 10161 (among others). They are basically small plates available on one side of a socket type USB (mini-B or Micro-B, as appropriate) where external power is turned on the other side of a JST 2 pin socket where connect the LiPo battery to be charged. They are prepared to be fed with an external source of 5 V; this voltage it can offer a USB wall charger any existing mobile or even a USB socket on your own computer. Thank you the controller chip MCP73831 these chips have a built this voltage

5 V is conveniently lowered to 3.7 V standards required by LiPo batteries. However, some of these battery chargers to provide a current of 500 mA, other than 1 A, etc., so we should know the capability of monitoring our bank for loader which model is best suited.

Similar to those described in the preceding paragraph may be found in other distributors listed in Appendix A and many other shippers. For example, in Adafruit they offer their product number 259, a USB / JST charger as above and the product No. 280, USB socket also incorporates a 2.1 mm female plug, allowing also obtain power from an operational external source between 5 V and 12 V. Thus, the product number 8293 is similar Sparkfun. Another plate that even allows you to charge LiPo batteries from solar panel via a pluggable JST is called "Lipo Rider" of Seeedstudio.

Characteristics of AC / DC adapters

The other type of external power source, different battery / batteries which most use to our circuits is the AC / DC adapter. Its typical feature is connected to an electricity grid of transforming the high alternating voltage offered by it (in Spain is 230 V + 5% and + 0.3% 50 Hz; if you want to know the other countries, you can consult http://kropla.com/electric2.htm)

in a continuous, constant and much lower voltage, to offer then this the apparatus to be connected and put into operation and a stable and secure manner.

The AC / DC adapters are basically formed by a transformer circuit which converts the AC voltage input to another AC voltage much lower, and a rectifier circuit which converts the AC voltage and transformed into a DC voltage which is the voltage output end. All adapters incorporate a printed label that informs both the range of values in the AC voltage input with which they are able to work (in addition to the frequency of the supported AC signal) as the value of the DC voltage and the maximum current They provide output. For example, the following image is an AC / DC adapter that supports AC input voltages between 100 V and 240 V at a frequency of 50 or 60Hz (therefore compatible with the Spanish grid) and provides a DC output voltage 9
V (and a maximum current of 1 A).

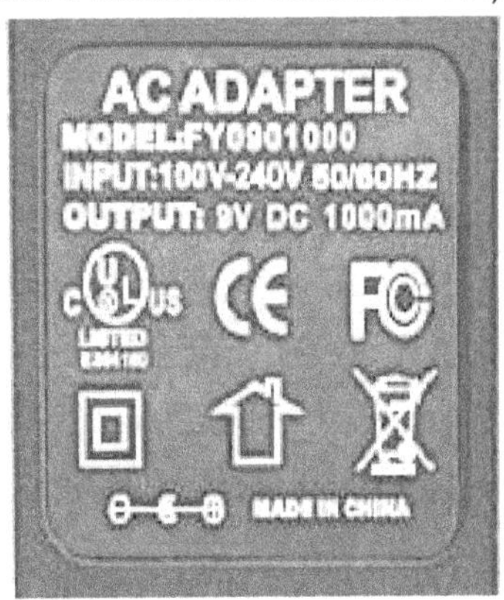

Adapters can be classified according to whether they are "regulated" (ie, if they incorporate a voltage regulator inside) or not. A voltage regulator is a device (or a set of them) which, being subjected to a certain relatively fluctuating input voltage, is capable of generating an output voltage normally lower, more stable, consistent and controlled.

Thus, regulated voltage adapters provide very specific and constant output which is equal to that shown on the label. What may vary (up also shown on the label) it is the intensity of available power, as this depends at all times on the needs of the feed circuit.

Unregulated, however, adapters do not have any stabilization mechanism and provide an output voltage whose value can be different in several volts to that shown on the label. Such adapters certainly reduce the input voltage to a lower output value, but the real value of this output voltage depends largely on electricity consumption (measured in amperes or milliamperes) performed at that particular time by the feed circuit.

Let us explain: as the circuit consumes more current, the output voltage (initially considerably higher than the nominal value marked on the label of the adapter) is reduced more and more to reach its nominal value only when the circuit consumes the maximum current that the adapter is able to offer, (whose value is also indicated in the printed label, as we know). If the circuit continues to increase its consumption and exceeds the maximum current, the voltage provided by the adapter will continue to decline and become less than the nominal, circumstances in which the risk of damaging the adapter (rebound and run the circuit fed). This behavior is easy to check with a multimeter, as we shall see in a later section of this chapter.

The main reason for the existence of unregulated adapters is their price: they are cheaper and are also available in a variety of shapes and ranges of use. Generally, the adapters that connect directly to the network plugs (as a "wall wart" or "wall-wart") are often unregulated. Those who, like those used in laptop computers, have a rectangular box leaving the cable plugged into the mains and connect the cable to the device, usually regulated. Anyway, regardless of encapsulated adapter, a rule is a rule often meet is that if the adapter supports a wide range of input voltage (100 to 240 V, for example), then surely it is regulated.

A (regulated) adapter that can come to us well for our projects Arduino is for example the product No. 63 of Adafruit: it is an adapter compatible with the Spanish grid, and generates an output voltage of 9 V and a maximum current of 1 A, which makes it perfectly compatible with Arduino boards (also its pin-type "jack" is 5.5 mm / 2.1 mm, as is the base of said plates). We could also serve 798. No product can even buy adapters that instead of the jack plug 5.5 mm / 2.1 mm, offer a USB connection type: an example would be the product code DFRobot FIT0197, which provides 5 V and a maximum current of 1 A.

If we need to change over output voltage (because we have our circuit devices that perform higher consumption, such as many types of engines) would have to use a (regulated) adapter such as Product No. 352 of Adafruit (which It provides an output voltage of 12 V and maximum current of 5 A) or No. 658 (5 V and 10 A). We must be careful, however, not to use an adapter that provides a voltage or intensity greater than the circuit output is able to admit: if we connect for example an Arduino latter two adapters, the plate was burned.

ELECTRICAL

Resistances

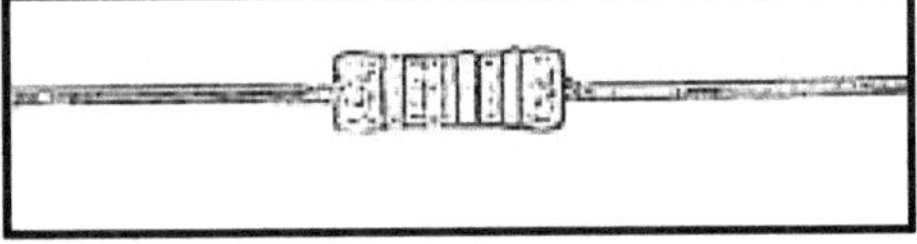

A resistor or resistance is simply an electronic component used to add, as the name suggests, an electric resistance between two points of a circuit. Thus, thanks to Ohm's Law, we can distribute as we suits different voltages and currents along our circuit.

Due to the small size of most of resistors, it is usually not possible to screen print the value of encapsulation, so to know we know how to interpret a series of colored lines arranged along its body. Typically, the number of colored lines is four, the last of golden or silver (although it can be other colors as well). This golden or silver line indicates the tolerance of the resistance, ie the accuracy of this brings us factory. If it is gold indicates a tolerance of + 5% if one silver + 10% (other colors-red, brown, etc. - indicate other values). For example, a resistance of 220 Ω with a silver tolerance band, may have a value between 198 Ω and 242 Ω (i.e., 220 Ω + 10%); Obviously, the lower the tolerance, the higher the price of resistance.

The three other colored lines indicate the nominal resistance value. To properly interpret these lines, we must place on our right the line of tolerance, and start reading from left to right, knowing that each color corresponds to a different digit (0 through 9). The first and the second line we take each digit such as which (one after the other) and the third line represents the number of zeros that must be added to the right of the last two digits. The table for the numerical significance of the possible colors of a resistor is:

Band Color	(in the 1st and 2nd band)	(value of the 3rd band)
Black	0	$x10^0$ =1
Brown	1	$x10^1$ =10
Red	2	$x10^2$ =100
Orange	3	$x10^3$ =1000
yelow	4	$x10^4$ =10000
Green	5	$x10^5$ =100000
Blue	6	$x10^6$ =1000000
Violet	7	$x10^7$ =10000000
Gre	8	$x10^8$ =100000000
White	9	$x10^9$ =1000000000

For example, if we have a resistance to the lines of "red-green-orange" colors, we can query the table to conclude that we will have a resistance
25 · 1000 = 25 KΩ. Another example: if we have a resistor with color lines
"Brown-black-blue," then we will have a resistance of 10 · 1000000 = 10 milliohms.

We can also find resistances that have five printed lines: in that case, his performance is exactly the same, only instead of two have three lines to indicate the first three digits of the resistance value, being the fourth which represents fifth multiplier and tolerance. Some even have resistances up to six printed lines (they are the most accurate, but our projects rarely the need); in that case, the only difference is that a sixth line to the right of the line of tolerance indicating a new fact: the temperature coefficient of resistance, which informs us about how much value that varies depending on resistance room temperature (measured in ppm / ° C, where 10000 ppm = 1%). Other resistances (especially those of small size, as the welded directly to the surface of a printed circuit board) used instead of colors, a sequence of three digits to indicate the first two figures of the resistance value and the multiplier.

It is clear that although to know the order of the stripes and read the resistance value we place this in a certain sense, the resistors have no polarity. This means that when connecting a circuit is indifferent to connect its two terminals in one direction or the other way around.

On the other hand, besides knowing the resistance value provided by these components, we must also take into account the current intensity that can support a maximum without melting. For this, the manufacturer must always provide us a fact: the maximum power that the resistance is able to dissipate as heat, which value is directly related to their size. The resistors used in most electronics are 1 / 4W 1 / 2W and 1W (being 1 / 4W is the smallest and 1W is the largest of the three). In this sense, to know what kind of resistance we use in our circuits interested, we use the already known formula $P = V \cdot I$ (where V is the potential difference between the ends of this resistance and I is the current flows through it) or one of the equivalent formulas, as P = I 2 · R or P = V2 / R (where R is the value of the actual resistance). In any of these ways, we get the power to be able to dissipate our strength, so that we will have the discretion to select it. Not a bad idea to use a resistor whose power sink is about twice the result to not suffer any overheating.

The symbols used in the design of electrical circuits to represent a resistor can be two:

where the right is the standard for the "International Electrotechnical Commission" (IEC) standard, although the left is still widely used today.

Potentiometers

A potentiometer is a variable resistance value. We can realize its great utility with a very simple example: if we assume a power supply that generates a certain stable voltage, and we remember Ohm's Law ($V = I \cdot R$), we can see that if we increase value resistor R, a voltage equal current intensity that passes through the circuit inevitably decrease. Conversely, if we decrease the value of R, the current I will increase. If this variation can control the R

us at will, can change as we like the current that flows through a

circuit. In fact, a very common use of potentiometers is to make progressive voltage dividers, so that we can, for instance, gradually turn on or turn off a light as we go changing the value of R.

A potentiometer physically has three pins: between the two ends there is always a fixed resistance value (maximum, in fact), and between either of these extremes and the center pin we have a part of that maximum. That is, the maximum resistance of the potentiometer between its two ends is simply the sum of the resistances between one end and the center pin (call it R1) and between the center pin and the other end (call R2). From here one can think that a potentiometer is equivalent to two resistors in series, but grace is that at any time we change the status of the center pin for increasing the resistance of R1 (slow because the resistor R2, as the maximum total value itself remains constant) or the opposite, to achieve lower resistance R1 (thereby increasing the resistance R2 automatically).

The concrete way to change the status of the center pin of the potentiometer can vary and often depend on their physical encapsulation, but usually usually involves moving a cursor manipulated connected to that pin. We may find potentiometers rotary motion as the volume control of most speakers, or rectilinear motion as those used in audio mixers, among others. In the image shown above you can see one of rotational movement.

There are also digital potentiometers type: these are chips comprising different pin through which electrical pulses can be controlled by the extreme values of the resistor and its intermediate value. One example is the Maxim DS1669 component.

The most interesting classification, however, comes to distinguishing the behavior that has a potentiometer when we change the status of your center pin (ie, when "is moved"). If the potentiometer has a behavior called "linear", altered resistance value is always directly proportional to the travel of the center pin: ie if the pin move eg 30% is increased / decreased by one resistance

30% also. Conversely, if the potentiometer has a "logarithmic" behavior, altered resistance value is initially very slight travel of the pin (and therefore will have to make a large displacement of this

for a significant change in resistance) but as you continue to make more travel of the pin, altering the resistance it will become proportionately larger and larger, until you reach a point where a slight displacement produce a large change in resistance . Logarithmic potentiometers are normally used for audio, since humans can not hear linearly: to experience such an acoustic feeling of "twice as hard", it is necessary that the physical volume of sound is about ten times higher.

The symbols can be used in electronic circuit design to represent a potentiometer they are:

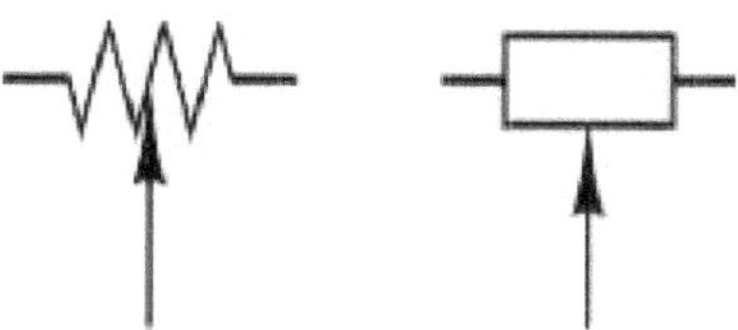

Other variable resistors value

The potentiometers are resistors that change their value according to our will. But there are resistors that change their value according to external environmental conditions.

For example, photoresists (also called LDRs English -from "Light Dependent Resistor" - or even "CdS cells" -for the material with which they are usually made: cadmium sulfide) are resistances that vary the amount of light strikes them, so they can be used as light sensors. Another example thermistors: resistors that change their value vary according to the ambient temperature, which can be used as temperature sensors. Another example is the force sensors / pressure (also called FSRs English -from "Force-Sensing Resistor" -), which are resistors whose value depends on the force / pressure to which they are subjected, or bending sensors: resistance whose value varies depending on what you are physically bent, etc.

In the chapter on sensors we will speak more extensively about the different characteristics of these components and their possible practical applications. If what you want is to know the schematic symbol corresponding to a particular type of variable resistor (LDR, thermistor, etc.), see the next page recommend

web, which includes most existing electronic symbols, categorized and sorted by categories: http://www.simbologia-electronica.com

Diodes and LEDs

The LED is an electronic component with two connection ends (or "terminals") that allows free passage of electric current in one direction only, blocking if the current flows in the

opposite direction. This makes the diode has two possible positions: a downstream (called "forward bias") or against ("reverse bias"). Therefore, when using it in our circuit, we must bear in mind that the connection of the two terminals is made in the desired direction. Typically, manufacturers will tell us what is the terminal to be connected to the negative (assuming forward bias) by a visible mark near this painted on the body of the diode. In the image shown, this is the thick white mark to the right of the body diode strip, so the "negative terminal" will, in this case the right. Technically (always assuming direct polarization) that "negative terminal" is called "cathode" and "positive terminal" is called "anode".

The diode can be used for many purposes: a common use is to rectifier (to convert alternating current to direct), but in our circuits we will use primarily as an additional element connected to any other part to prevent this damage if power is mistakenly connected with the polarity reversed.

It is customary to connect a voltage divider (that is, a series resistor) to one of the diode (it is immaterial whether the anode or cathode) to prevent either the precisely which case the more tension you can support. To calculate the value of this resistance, we must consider the intensity that must pass through the diode (I), the voltage across its terminals that would exist if we did not put up any resistance (V) and the voltage across its terminals we want to achieve to avoid damage (VDIO); once you are known these values we can calculate the proper resistance using Ohm's Law, as follows: R = (V-VDIO) / I. Furthermore, the power dissipated by this resistance could calculate using the formula P = (V-VDIO) · I.

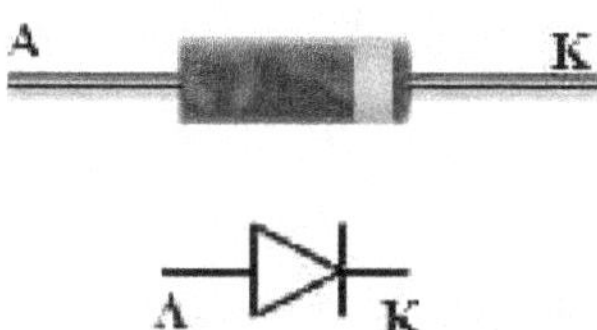

A "Light Emitting Diode" (LED) is, as its name suggests, a diode that has a peculiar feature: emits light when electrical current passes through it. In fact, it does in proportion: the more intensity current passing through it, emits more light.

Since it is still a particular type of diode can also be connected directly or reverse bias, considering that only
will illuminate if connected in forward bias. Therefore, when designing

our circuits must continue taking care to connect each LED terminal in the proper polarity. However, as an LED does not he can paint a mark-up, how to distinguish the anode ("positive terminal" in forward bias) of the cathode (the "negative terminal" in forward bias) is observing its length: the anode is a longer length than the cathode.

It is also highly recommended to connect a resistor in series with an LED to limit the intensity of current passing through it and keep it well below the maximum value beyond which the LED may be damaged. To calculate what resistance we place, we can use the formula mentioned in a few paragraphs above. We know that normally the intensity that often come in handy for optimum performance of an LED is about 15 mA, and that the proper voltage varies VDIO LED color: ranging from 3 V to 3.6 V for the ultraviolet (UV), white or blue, 2.5 V to 3 V for green, 1.9 V to 2.4 V for red, orange, yellow or amber and 1 V to 1.5 for the infrared. From here, the calculation is simple. However, for safety it is advisable to always consult the specifications provided by the manufacturer (the "datasheet" component) for all information on currents and maximum voltages supported.

Apart from its different colors (resulting from manufacturing material used, different for each type of LED), we can classify these components as if they emit light in a diffuse or clear. The first (which usually have a size of 3 mm in diameter) are usually used to indicate presence, giving off an even soft light that does not dazzle and can look good from any angle. The second (which usually have a size of 5 mm diameter) are used to radiate in a very specific direction direct and strong light, so it does not look good at all angles but much more light than others. In any case, it is a diffuse LED or light type, for quantitatively know what "bright" is his light, it is necessary to know the amount of millicandelas (LCD) that this particular LED is capable of emitting; this data it should be offered the manufacturer in the datasheet.

Finally, we must consider at least three issues when using diodes and LEDs in our projects: the maximum current that can pass through the diode forward bias is based without this due to heat generated by the power dissipated; the breakdown voltage (when diodes are connected in reverse bias have said do not let the current flow but such behavior is so only as long as the diode will apply a lower voltage of the "breakdown voltage") and fall forward bias voltage (such as other electronic devices, diodes having an internal resistance caused by the existence of a certain potential difference between its terminals).

The symbol is often used in the design of electronic circuits to represent a standard diode is shown on the left in the picture below, and an LED is to the right.

The apex of the triangle with the secant line perpendicular both symbols represents the cathode. There are other similar symbols that represent more specific diodes (such as Zener or Schottky type, among others) but we'll see.

Capacitors

The capacitor is a component whose primary function is to store electric charge in limited quantities, so that this can be used in very specific occasions as an "alternative

power source."

The capacity (C) of a capacitor is its most important feature and can be defined as the ratio-usually a value constantly between the amount of electric charge (Q) stored at a particular time and voltage (V) that is being applied at that moment. Specifically, it is defined as: $C = Q / V$.

From the above formula we can deduce several things: the first is that a capacitor with larger capacity than other store more charge under the same potential. The second is that a capacitor with a given load capacity store more the higher the applied voltage (although in this respect it should be noted that all capacitor has a maximum working voltage

which often come printed on the body of the condenser-own beyond which is

It can damage, so always maximum load will be stored).

Capacity is measured in farads (F), although most will work with capacitors have much lower capacity (about the microfarads or even nanofarad). Depending on the size of the capacitor, it may be that the value of his ability can not be screen-printed as is on your body; in such cases, it is often used a sequence of three digits to indicate the first two digits of the value of the capacity and then your multiplier. For example, a capacitor with the number "403" printed, it means having a capacity of

$103 = 40 \cdot 40 \cdot 1000 = 40000$ F.

As was the case with resistors, capacitors may be connected in series or in parallel to achieve a circuit with an equivalent capacity. Specifically, if two capacitors (C1 and C2) are connected in series, the equivalent capacitance is $C = (C1 \cdot C2) / (C1 + C2)$; ie, less than any of the individual capacitors. If the two capacitors are connected in parallel, the equivalent capacitance is $C = C1 + C2$; that is, we obtain a higher total capacity.

A fully charged in DC circuits, capacitor acts as an open switch.

We can classify the types of capacitors according to whether they have polarization. Polarized capacitors are those that are to be connected to the circuit observing the direction of the current. That is, have a "negative" terminal must always be connected to the negative pole of the circuit, and a "positive" terminal must always be connected to the positive pole. In other words: the connection must always be carried out in forward bias. That is why this type of capacitors are invalid

for use on alternating current; in fact, when connected in reverse polarity are destroyed. To distinguish one from the other terminal, we can look at a strip painted on the body of the capacitor, which always indicate the terminal negative (similar to what happened with the diodes). Another hint that we can use for the same purpose is to look at the negative terminal is shorter than the positive (as was the case with LEDs). The most outstanding features, we can say that the polarized capacitors have as a rule a fairly high capacity (of a value greater than 1 microfarad), and taking into account the manufacturing, usually electrolytic type (as shown in the picture) or tantalum.

Unipolar capacitors (not polarized) can connect to the circuit indifferently either way (as resistors, for example). They can be made of many materials, but the most common are the ceramic type (as shown in the picture). Usually typically they have a much smaller capacity than the polarized capacitors.

Capacitors are often used in circuits to provide the "power deviation" or "decoupling" (in English, "bypass" or "decoupling"). This power is required when a component that usually do not require much amperage to run, you have to make a timely manner so high electricity consumption than ordinary power source is not able to provide it quickly enough. This normally happens when said component (a microcontroller, for example) switches from off to on in state; at that time it is when the capacitor provides fast transient current required to "drop" the electric charge that kept stored. With this system of power "alternative", it is possible to give a quick response to peak consumption of the component while the power supply can go again reloading the condenser (at a slower pace) for next peak. To use a capacitor to perform this function, one of the terminals should be connected as close as possible to the power input of the component to "manage" (the farther, the smaller the effect) and the other terminal to circuit ground . Typical values of a capacitor "bypass" or 0,01µ F are 0.1 uF.

Another common use of capacitors is to eliminate the "noise" of the DC power signal. That is, although a source is nominally labeled as 9 V, for example, in reality never offer those exact 9 V, but that value will suffer more or less wide variations and
random around its nominal value. Therefore, in addition to a DC power supply always we have a small AC around that. Depending on the magnitude of the AC power (ie, variations around 9
Example V), we might have a problem in our circuit, that appear effects of type AC currents do not want. A capacitor is able to stabilize these variations, allowing to obtain a value of the source of constant tension, thanks to its ability to properly regulate charging that "loose" or "accumulate" depending on the fluctuating voltage to which it is subjected.

To use a capacitor to perform this stabilizing function of the signal (which are often called "filter capacitors"), one of its terminals must be connected to the positive terminal of the power supply and the other terminal must be connected the negative terminal. A typical value of a filter capacitor is 0.1 uF. A higher frequency AC capacitors smaller capacity are needed. You can even connect multiple capacitors in parallel with different capacities to filter different frequencies (but the actual calculation is beyond the scope of this book).

Actually, the capacitors are used in many applications: as batteries and memories for its quality of stored charge for fast downloads (as light "flash" of a camera) to maintain stable currents (such as those generated by a rectifier) to prevent falling point current circuits (ie, the function of "by-pass"), to isolate parts of a circuit (when fully loaded), etc.

The symbols used in the design of electrical circuits to represent a capacitor may be two:

The representation of the left is that of a unipolar capacitor, and the right is of a polarized capacitor. In the latter, the straight line symbolizes the (sometimes the "+" is omitted and / or the straight line thicker paints) positive pole and the curve symbolizes the negative pole.

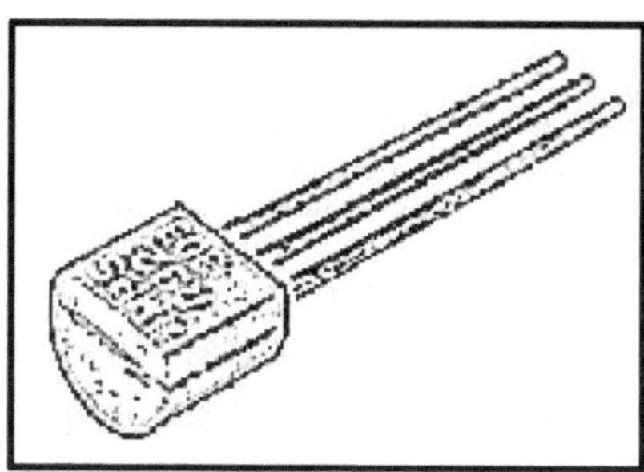

A transistor is an electronic device that restricts or allows the flow of electrical current between two contacts according to the presence or absence of current on a third. It can be understood as a variable resistance between two points, the value of which is controlled by applying a certain current on a third point.

Transistors are often used as power amplifiers, because with a small stream received through its control terminal allowing the circulation of a very large intensity (proportional to that, up) between its two output terminals. Another common use is to be current switches, because if your control terminal receives no current, by between two output terminals either no current flows and the circuit is opened.

There are two broad categories of transistors by manufacturing technology and operation: bipolar transistors (BJT commonly called, the English "Bipolar Junction Transistor") type transistors and field-effect (FET commonly called, the English "Field Effect Transistor ").

The BJT transistors have three physical pin and each has a specific name: "Collector", "Base" and "Issuer". The "Base" makes "terminal control" and "Collector" and the "Issuer" are the "output terminals". However, depending on how they are used and connect these three pins, we can classify the BJT transistors turn into two types-the most ordinary work NPN and PNP. In the case of NPN, if we apply certain current (usually very low) Base the Issuer, the Issuer will act as a valve to regulate the flow of current from the collector to the emitter itself. In the case of the PNP, if we apply certain current (usually very low) of the emitter to the base, the Issuer will act as a "valve" that regulates the current flow from the emitter to the collector himself.

Physically, BJT transistors can be very different, but the one shown in the figure on the previous page is a typical encapsulated, in which three pins correspond to the collector, base and emitter (although to tell which is which is
should consult the device datasheet). The symbols used in the design of electrical circuits to represent the NPN and PNP transistors are:

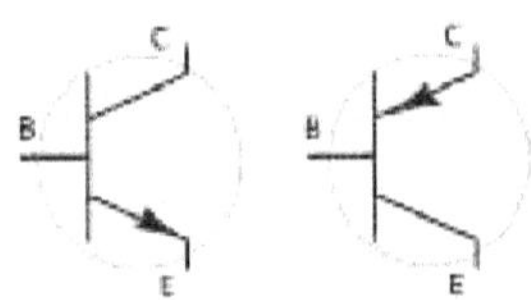

When the transistor is used as an amplifier, the intensity of current flowing from the emitter to the collector (if PNP) or collector to emitter (if NPN) can be tens of times larger than the current apply to the input terminal Base-Emitter. This proportionality factor between the current passing through the base and passing through the collector (that is, the amplification factor or gain) dependent transistor and is to come described in the technical specifications (the "datasheet") with the name or β hFE.

More specifically, we can distinguish three "modes" in a typical transistor:

The cutting mode: occurs when the current flowing through the Base is close to 0. In this case, no current flows through the inside of the transistor, thereby preventing current flow both the Issuer and the collector (ie , the transistor behaves as an open switch).

Saturation mode: it occurs when the current flowing in the collector is virtually identical to that flowing through the emitter (at which, in fact, these approach the maximum current value that can withstand the transistor itself) . That is, the transistor behaves as a simple wire bonding, since the potential difference between collector and emitter is close to zero. This mode occurs when the current flowing through the base exceeds a certain threshold.

Active mode: occurs when the transistor is neither in its cutting mode or in its saturation mode (ie, in an intermediate mode). It is in this mode when the current flowing through the collector current depends mainly on Base and β (current gain). Specifically, it holds that $Ic = \beta \cdot Ib$ and $Ie = Ic + Ib$, where Ic is the current flowing through the collector, Ib which flows through the base and the Ie flowing by the Issuer.

As you can see, the active mode is interesting to use the transistor as a signal amplifier, and cutting modes and saturation are interesting to use the transistor as a switch that represents the low and high logic state respectively.

For its part, the FET transistors perform the same function as the BJT (amplifier or switch current, etc.), but its three terminals are named (rather than Base, emitter and collector): Door (identified as "G" , English "Gate"), Supplier (S) and drain (D). The G terminal would be "equivalent" to the Base in BJT, but the difference is that the G terminal does not absorb current at all (compared to BJT where the current flows through the base despite being small compared to the circulating by the other terminals). The terminal G rather acts as a voltage controlled switch, because (and here's the key operation of this type of transistors) will be the existing voltage between G and S which allows current to flow or not between S and D.

As well as BJT transistors are divided in NPN and PNP, the field effect (FET) are also of two types: the "channel n" and "channel p" depending on whether the application of a positive voltage the door puts the transistor in conduction or non-conduction, respectively.

Moreover, the FET transistors can also be classified in turn depending on its structure and internal composition. Thus we have the JFET (Junction FET), the MOSFET (Metal-Oxide-Semiconductor FET) or MIS-FET (Metal-Insulator-Semiconductor FET), among other transistors. Each of these types has different specific characteristics that make them more or less advantageous depending on the needs of the circuit, and each has a different schematic symbol.

In general, the FET transistors are usually used more than BJT circuits that consume large amounts of power.

Buttons

We know that a switch is a device with two physical locations: in the position "closed" connecting two terminals (allowing current to flow through it) occurs and in the position "open" occurs disconnection of two terminals (and hence the current flow is cut through it). In short, a switch is simply a mechanism comprising a pair of electrical contacts are bonded or mechanically separated.

A button (in English, "pushbutton") is simply a type of switch in the on position which is set by the push of a button thanks to the pressure exerted on a conducting internal sheet. At the time of leaving the press on this button, a spring causes the sheet to regain its former position, returning to the position "open".

A variety of buttons of many shapes and sizes, but the circuits that will take place throughout this book we will use very few buttons
practical and relatively small (having a size of only 1/4 inch
per side) that can be purchased from any distributor listed in Appendix A (eg Sparkfun it is the product No. 97) and which are usually come in kits learning.

These buttons, however, have a feature that should be clarified: as shown in the side image, have four pins. This may make us think we have four terminals, but nothing is further from reality: the two pins facing each side are connected internally to each other, so they work as one. That is, pins A and C highlighted in the image represent a single electrical point, and pins B and D represent another single electrical point. So really, this button only has two terminals: A / C and B / D.

In the next picture you can see the power button symbol in (left and right open state in the closed state), also indicating the example of what physical pin corresponds each terminal.

Keep in mind that this type of buttons usually have quite limited maximum voltage and current amount that can withstand before burning: If you are going to use a more demanding power, they must use more robust switches. For the specific values of voltage, current (and pressure on the button, among other things) maximum that can withstand a particular button, we have to refer to the official documentation provided by the manufacturer for that component (the "datasheet").

Voltage regulators

A voltage regulator is an electronic component that protects parts of a circuit (or an entire circuit) of high voltages or variations of this pronounced. Its function is to provide, from a fluctuating voltage received within a certain range (so-called "input voltage"), another voltage (called "voltage output") set to a value less stable. This can be achieved by applying Ohm's Law: thanks to its ability to increase or decrease (as appropriate) the internal current that is going through
each time, they can raise or lower its output voltage proportionally.
So I explained it might seem a simple voltage divider (and indeed, their function is the same), but its mechanism for regulating the output voltage is much more sophisticated and reliable.

They are a key to a correct and safe of the various electronic components of our supply circuit element. Thanks to them, components that would be damaged if subjected to too high a voltage, can be combined into a single circuit with more capable of components and requiring a higher voltage.
There are many types and models of regulator, but in our circuits typically use smaller components called generically encapsulated LDO regulators (for "low-dropout"). The voltage "dropout" is the difference between the input voltage and output. This voltage multiplied by the current through it is wasted as heat, so the lower the dropout, the more efficient the regulator in terms of energy loss. For example, if we have a LM8705 regulator fed to 12 V which provides a regulated voltage of 5 V and it is estimated that the maximum current that our circuit is 1.2 A, can calculate the power lost as heat by the regulator with well-known formula P = V · I: (12 V-5V) · 1,2ª = 8.4 W. The value obtained is quite high, but fortunately, is the worst because we considered the maximum current drawn by the circuit; microcontrollers in general and many electronic devices consume current pulses, so the average current supplied by the regulator is usually much lower, and therefore, much power (both cold) in the same is not lost.

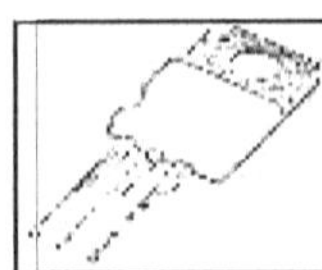

The LDO regulators usually have three pins: one for receiving the input voltage, the other to provide the output voltage (which would "positive terminal" for sensitive components) and a third pin connected to the common ground with the power supply . But the order and location of each pin depends on the particular model of controller, so you should consult the manufacturer's technical documentation to meet her.

The family of LDO regulators most widely used in home electronics projects is the LM78XX, where "XX" indicates the output voltage. So many times the specific model that will interest us is the LM7805, which can receive up between 7 V and 35 V input and can generate a maximum output current of 1 A. A similar model is the LM2940. If we get a maximum intensity of 0.1 A, we then use the LM78L05 model. If we want an output voltage of 3.3, we can use the LM7803 or LD1117V33, among others.

Whatever regulatory model, most often we see attached to your entry earth pin-and a capacitor "bypass" and see connected to the earth pin-and output filter capacitor. The reason is to eliminate possible fluctuations in the input signal (for example due to the sudden ignition of an element of high consumption of our circuit, such as a motor) and output (voltage obtained thus optimizing the regulator). Therefore the most common connection scheme of a typical regulator (such as LM7805) is similar to this:

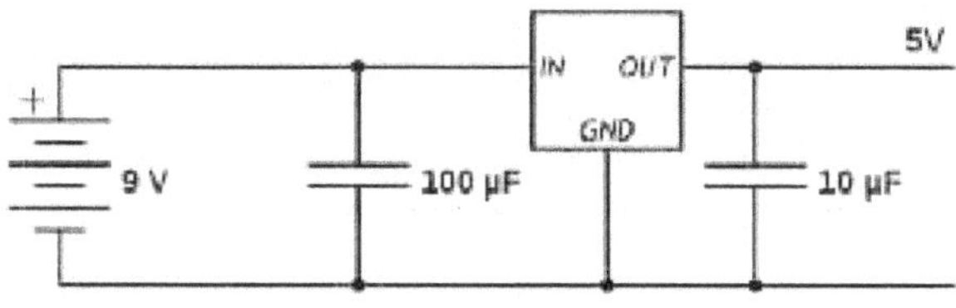

In the figure above you can see that we used a capacitor "bypass" of 100 microfarads and capacitor filter 10 microfarads. These values usually come well in most circumstances and are quite "standard". They could connect additional capacitors in parallel on both sides of different capacities to respond to major changes in input voltage and output respectively, but usually we do not find ourselves in situations where this is necessary.

Another common model is the LM317 regulator, whose most interesting feature is that you can adjust the output voltage to which we desire (specifically between 1.25 V and 37 V, with a minimum output current of 1.5 V) and that is capable of withstanding input voltages between 3 V and 40 V. To achieve vary the output voltage, have a connected regulator formed by a fixed resistor (R1) and a potentiometer (R2), the auxiliary circuit follows:

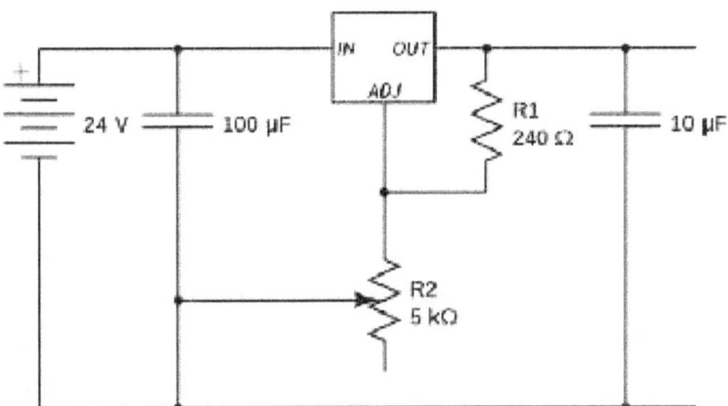

LM317 and using the circuit shown in the above scheme, obtain an output voltage given by the expression Vout = 1.25 · (1 + R2 / R1). Optionally, the previous design can be added a diode to protect the regulator against short circuits at its input; To do this, we should connect the anode of the diode to the output pin and the cathode to the input pin.

Prototyping boards

There are several types of prototyping plates. This section only we study called "breadboards" (also known as "breadboard"), the "perfboards" and "stripboards".

A breadboard is a perforated plate with internal connections where we can insert our paws electronic components many times as we want, thus realizing our circuits connections without the need to solder anything. The goal is to ride fast but fully functional prototypes of our designs and they can be easily modified when needed. The following figure shows the external appearance of a typical breadboard, which continues to be a set of rows with holes.

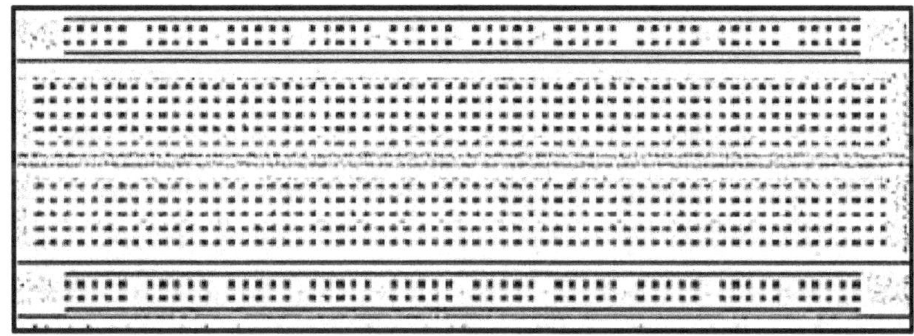

But in order to successfully connect our components to the breadboard, we must first know how to structure their own internal connections. Here, if observing inside hidden beneath the perforated surface, we could check that is composed of many metal strips (usually copper) arranged as follows:

In the previous figure we can distinguish basically three areas:

Buses: Buses are located on one or both sides of the breadboard. There will be connected (at any point) external power sources. Normally it is painted a red line that is often used to indicate the bus subjected to input voltage (ie, where insert the positive terminal of the source) and a blue line representing the bus grounded (ie, where normally insert negative) terminal. All points marked with red bus line are equivalent because they are connected together and all points on the bus marked with blue line are also each other, but both buses are electrically isolated from each other.

Nodes in the middle of the breadboard lot of holes appear. Their amount may be higher or lower depending on the model (in fact, the size of the breadboard is indicated by the number of rows and columns of holes comprising: a typical size is 10x64). These holes are used to position the components and make the connections between them. As can be seen in the previous figure, the internal connections between the holes are arranged vertically. The most important thing is to understand that any hole is completely equivalent to another belonging to the same internal connection. This means that when you insert a pin of a component in a hole, we have the other holes of the same internal connection to insert a pin in them any other component we want to contact each other, as if directly uniéramos a cable. To all those connected between holes equivalents are given the overall name of "node". The most common way to connect two different nodes is plugging the ends of a cable in a hole of each node to be joined.

Center channel: is the region located in the middle of the breadboard, which separates the top of the bottom. It is often used to place (such components in the form of "black cockroach with legs" also called "chips" or IC-from English "integrated circuits" -) integrated circuits so we put half of legs on one side of the channel and the other half on the other side. Thus, in addition to having several holes and connection for each leg, one half of the chip is electrically isolated from each other (as they should).

It is also frequently used "minibreadboards" specially designed for more compact projects because they lack the power and ground buses and its dimensions are smaller (thus accommodating fewer nodes).

In addition to knowledge of the internal electrical layout of a breadboard, it is important to consider a number of useful tips for the day to come us well when riding our designs. For starters, you should always use black cable to the grounding, red cable feeds of 5 V or more and green cable (or any other color) to 3 V feeds (and thus avoid damaging any component that does not support 5 volts). These colors are simply a general convention (ie, are not an established standard) but it is very common worldwide follow to avoid confusion.

Another tip is to see if our breadboard has painted the red and blue of its bus lines not continuously. If so, it means that the points that form each bus are not connected electrically all together but there is a jump. To splice all points of each bus and thus achieve a single power bus and a single ground bus (advisable to gain clarity and comfort in the realization of our circuits measure), what to do is one attached by a cable each end point jump for the power bus and joined by another cable each end point of the jump to the ground bus (perform what is called a "bridge").

On the other hand, a basic precaution must always be taken into account when using the ground bus is to ensure that all grounding circuit are connected in turn to each other for all the circuit has the same reference (which is sometimes called "sharing the masses"). That is, there is only a single ground for all components. This is critical for our circuits work properly.

For breadboards that have the power and ground buses on both sides (as there breadboads with buses available only in one) we can join the bus power on one side and the other by a cable and bus aside land and other means of another cable. This will help us to lead both pairs of buses running to connect a power supply, thus being easier and orderly handling circuit to ride.

Finally, remember that it is very important that when you add, remove or change components on a breadboard is not receiving any power. If you do not do this, you run the risk of shock and / or damage components. It is also important to check that the metal parts of the wires (or other components) do not contact each other because this would cause a short circuit.

Furthermore, besides the breadboards (breadboard), other types of prototyping plates, of which the "perfboards" and "stripboards" are the most important:

Perfboards: serve the same function as breadboards, but they get the prototype circuit more solid. They basically consist of a rigid, thin plate filled with pre-drilled holes in a grid and spaced at a standard distance. These holes must be welded us our circuit components and connections

among these are performed with cables we also welded to the plate. More specifically, the components are placed above the upper face of the plate (so their legs pass through its holes being welded these to the bottom face) and the wires are welded on the lower face (thus remaining relatively hidden the final assembly of the circuit). The underside of a "perfboard" the watching can easily identify the presence of copper rings surrounding the holes.

Stripboards: also known by the trade name of "Veroboards") are very similar to those perfboards. The major difference is that the holes of perfboard are electrically isolated from each other (and therefore have always making connections "manually") but the holes of one side of a stripboard are input linked by lines copper conductor, as separate parallel rows. In this sense, the stripboards are similar to those breadboards, as they already come with a number of holes -set nodes connected predefined himself. The only welds to be performed are therefore those of the components themselves and the cables connecting the various nodes.

Because welding skills are required (although minimal) in the projects in this book we do not use or perfboards or stripboards. However, if you want to learn how to use a "perfboard" a good tutorial for this it is http://itp.nyu.edu/physcomp/Tutorials/SolderingAPerfBoard. If you want to learn to use a "stripboard" we can see a good tutorial in the direction
http://www.kpsec.freeuk.com/stripbd.htm.

USING A prototyping board

This section shows some examples collected show different ways to connect devices using a "breadboard".

Example No. 1: In the diagram below you can see how a circuit where a resistor and LED are connected in series is performed

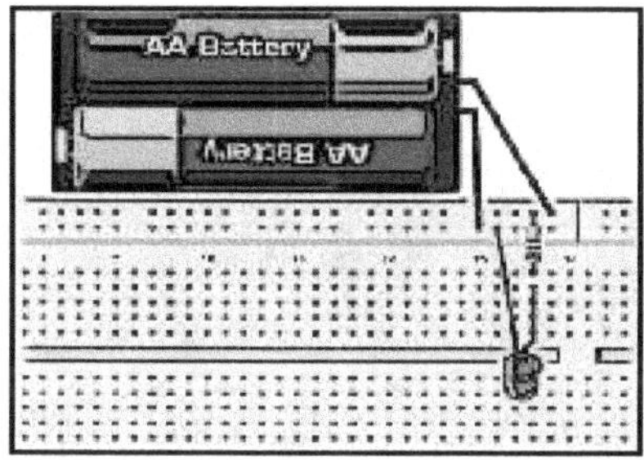

In the drawing above the positive terminal of the power supply is connected to the outermost bus. This will cause any element connected to a hole that you receive directly from there bus power. That is precisely what happens to the strength of our circuit: its upper end is plugged into the bus. The other terminal is placed on a breadboard node, node where precisely the positive terminal of the LED is also connected (in the diagram, is shown with a small slit at the root). This means that these two components are directly connected. Finally, the negative terminal of the LED is plugged into the internal bus of the breadboard, where bus also connects the negative pole of the source, so the LED is directly connected to it, closing the circle. The above circuit has a scheme like this:

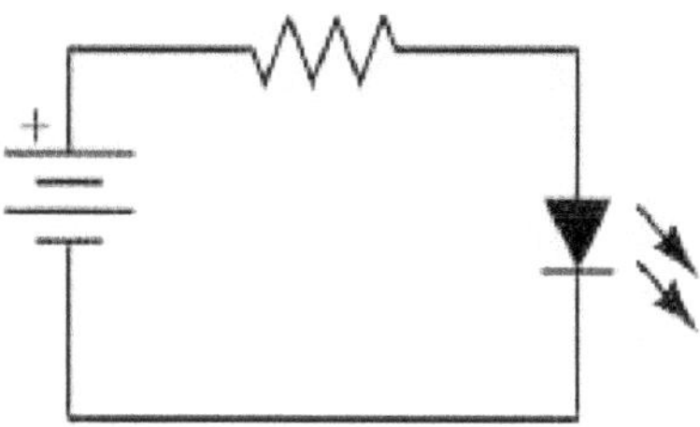

It could have also done the same circuit carrying the power and ground to the area of nodes:

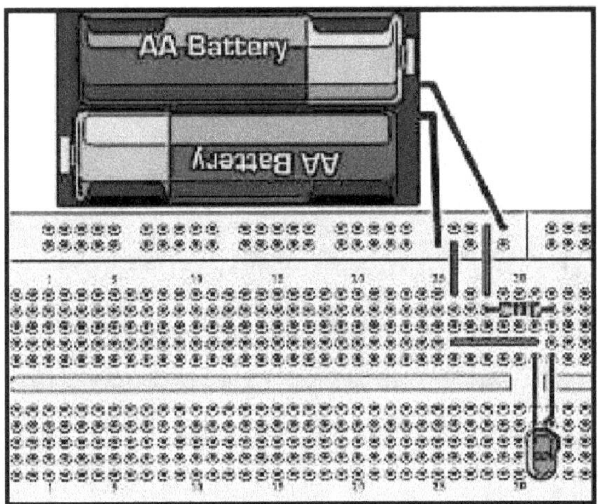

It is important to note that all holes of the same node represent a single point of connection (a very common mistake in the beginning is to connect both terminals of a device on the same node, which makes no sense). Knowing this, it is easy to see from the above illustration that the left terminal of the resistor is connected to a cable which in turn is connected to power, and the right terminal of the resistor is connected to the positive terminal of the LED while its negative terminal is connected to a cable which in turn is connected to another, which will eventually end up in land.

Example No. 2: The following illustration shows the connection of three devices in parallel (ie, three resistors):

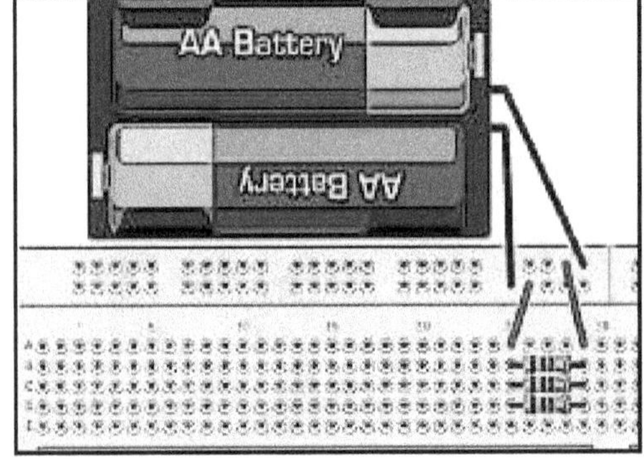

Equivalent scheme would:

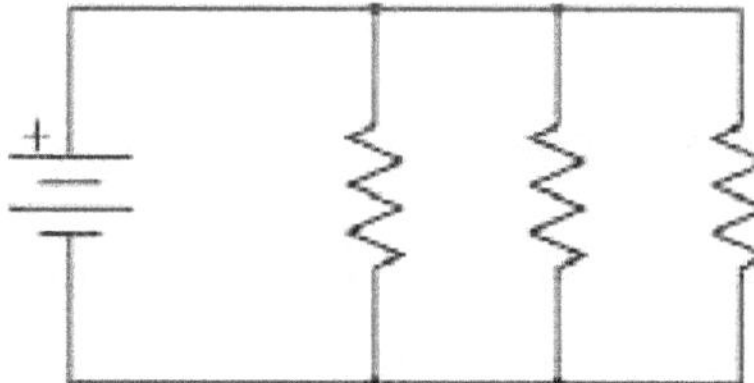

Example No. 3: The following diagram shows the series connection of a resistor and LED, and parallel connection of both LED and other resistance. If we try this circuit in reality, we see that depending on the value of each resistor, the corresponding LED will light more or less.

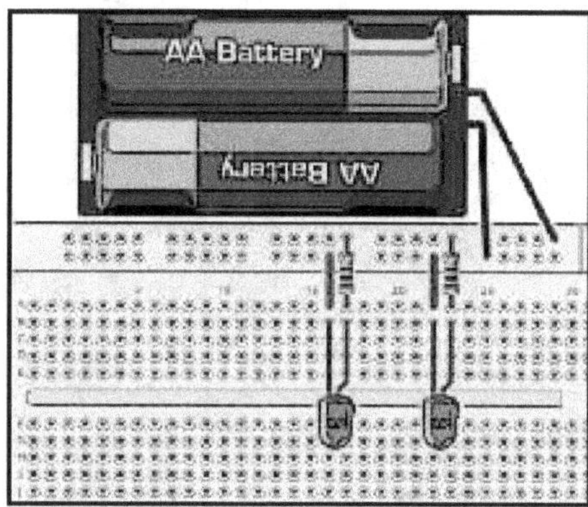

The above circuit has a scheme like this:

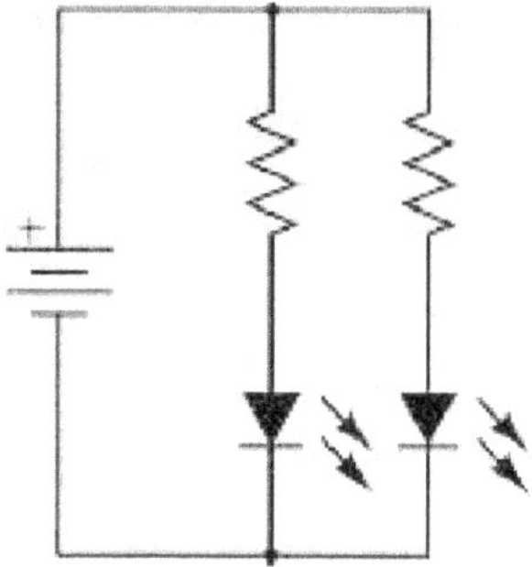

Example No. 4: The following illustration shows the connection in series of three devices: an LED, a resistor and a button:

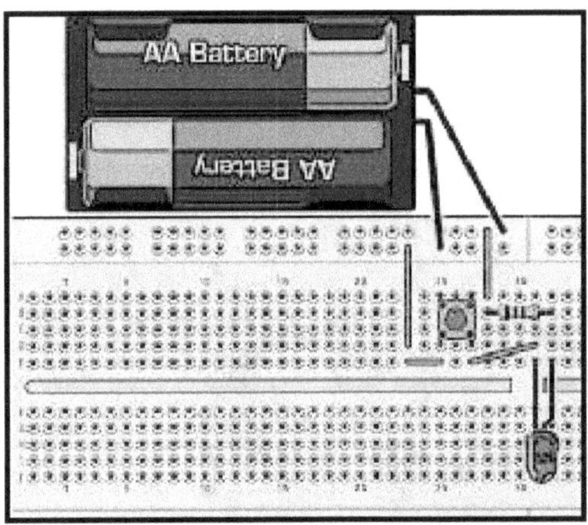

Also it could have done much the same circuit without wire:

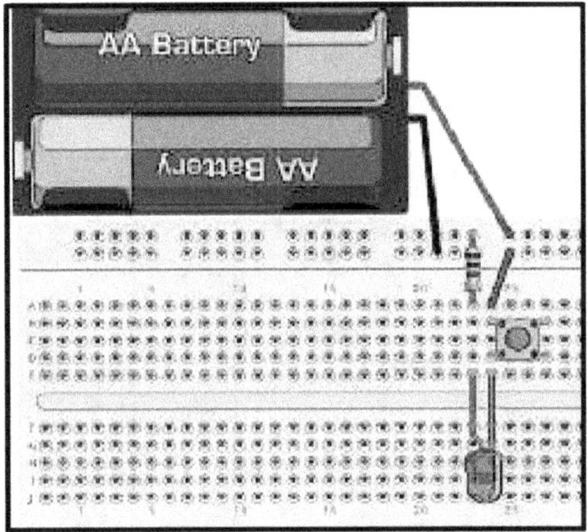

Please note that the above figures look at how the pins are located pushbutton. If on the last circuit had placed the button in the manner shown in the following figure, the LED would be on at all times because, regardless of the state of the button, the two pins facing are always connected internally (as in reality, represent a unique connection point, as we have already studied).

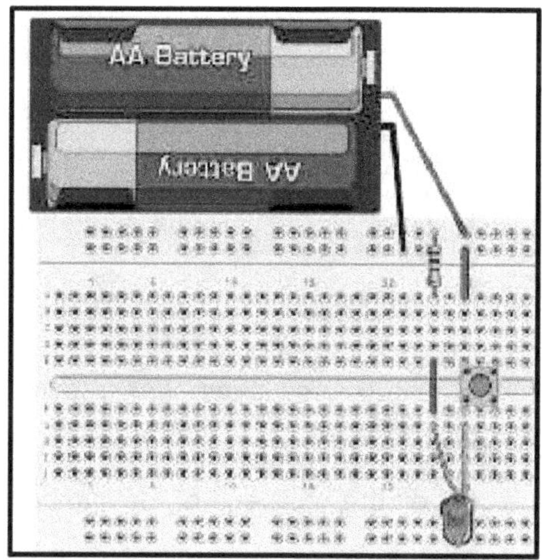

Example No. 5: The following diagram shows the connection in series of three devices: an LED, a resistor and a potentiometer:

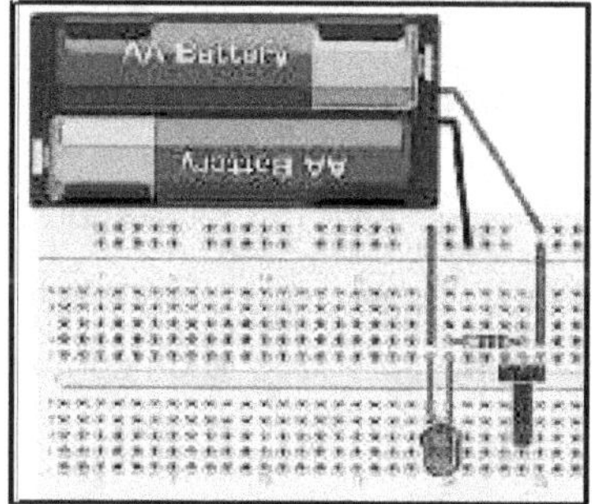

His scheme is the corresponding (where we can clearly see that the center pin of the potentiometer is connected and one end but not the other):

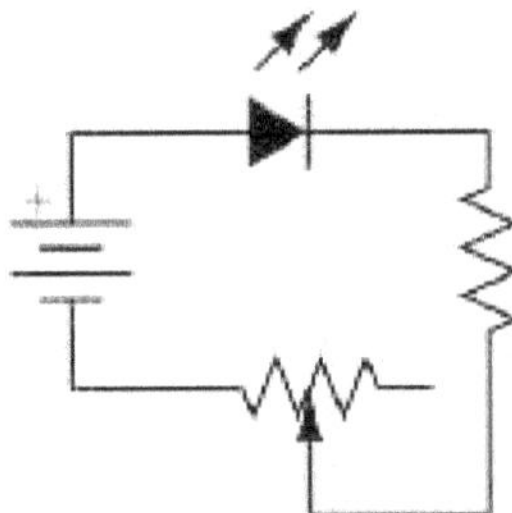

We have to imagine that the central arrow symbol potentiometer moves from one end to the other of that symbol as we turn the potentiometer screw. As the above diagram is drawn, if the arrow is "placed" on the right end of the symbol, the potentiometer function with maximum resistance value; if the arrow is "placed" on the far left, the potentiometer will not offer any resistance.

This last fact is the result of having added a resistance between the LED and the potentiometer: if at any time the potentiometer adjusted it to zero, there would be no resistance in this circuit !, and irreversibly damage the LED that receive too much amperage . Therefore, the additional resistance maintains a minimum value which is never lowered.

Finally, and on the other hand, if you want to feed a mounted circuit on a breadboard by an AC / DC adapter instead of batteries / cells (as shown in the examples above), we can use a few small plates especially intended to be connected on one side to pin

5.5 / 2.1 mm of said adapter (thanks to incorporating relevant socket) and the other to the breadboard. Moreover, these inserts contains a voltage regulator itself allowing adjustment of the voltage received adapter, most suitable for the type of circuits are usually mounted on a breadboard recessed and stable voltage (usually 5 V or 3.3 V). Examples of these inserts (among many others available in the various distributors listed in Appendix A) are Adafruit product # 184, # 114 (also the No. 10804) Sparkfun, called the "5 V / 3.3 Breadboard V DC-DC Power Supply "(and also" Breadboard 5V USB Power Supply ", which also includes a mini-B USB socket for power there too) of Akafugu or" Breadboard Power Supply Module "of IteadStudio (also with socket USB mini-B). In any case, you should consult the official documentation of each product to know thoroughly its features and limitations.

Use of a digital multimeter

The digital multimeter is an instrument used to measure any of the three variables related by Ohm's Law: either the existing voltage between two points in a circuit, or the intensity of current flowing through it, or the resistance of a component. Depending on the model, there are multimeters that can measure other quantities as the capacitor capacitance, and more. It is, it is a tool that allows us to verify the correct functioning of electronic components and circuits, so it is important to have it on hand when we make our projects.

There are many different models of digital multimeters, so it is important to read the manufacturer's manual to ensure the proper functioning of the instrument. An example might be the SparkFun Product No. 9141. However, although depending on the model can change the position of its elements and the number of functions, in general, we can identify the parts and functions of a generic standard multimeter as follows:

Button "power" (off-on): Most multimeters are powered by batteries.

Display: LCD where measurement results are displayed.

Set the key: used to choose the type of measured variable and the measuring range. The symbols that surround indicate the type of measured variable, and the most common are the forward voltage (V) and AC (V ~), direct current (A-) and alternating (A ~), resistance (Ω), the capacity (F) or frequency (Hz). The numbers indicated around the key measurement range. To understand the latter, assume that the possible numbers for continuous voltage are for example "200 mV", "2 V" "20 V" and "200 V"; This will mean that in the "200 mV" position can be measured voltages from 0 to the maximum value; in the "2 V" position can be measured above 200mV but below 2 V voltages; in the "20 V" position may measure greater than 2 V but below 20 V voltage, and thus the display showing the measured numerical values according to the chosen scale.

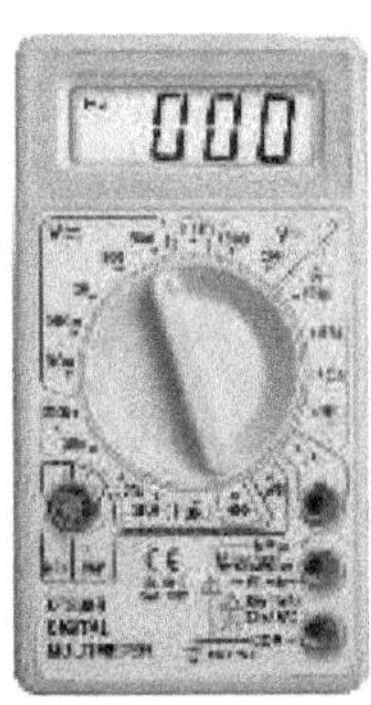

Red and black-tipped wires: the black wire always connect the black multimeter socket (there is only one, and is generally marked with the word "COM" -from "common reference" -), while the red wire connects to red socket warranted by the magnitude to be measured (as there are several): if you want to measure voltage, resistance and frequency (both continuous and alternate), it should connect the red cable to the red socket usually marked with the symbol " + VΩ Hz "; if you want to measure current intensity (both AC and DC), you should connect the red cable to the red socket marked with the symbol "mA" or "A", depending on the range to be measured.

Once we know the functional parts of this tool we can use to perform different actions:

To measure the voltage existing (continuous) between two points of a circuit fed, we conveniently connect cables to the meter to then place the tip of the black cable to a circuit point and the red wire to the other (so that we are actually making a connection in parallel with the circuit). Then we will move the selector switch to choose the symbol V and suitable measuring range. If you know this, we can do is start at the highest range and work down step by step to finally get the desired accuracy. If we go over the account (ie, if the value measured is greater than the range chosen), we'll know because the left of the display to the special value "1" is displayed.

We may also use the opportunity provided by the multimeter to measure DC voltage to meet the potential difference on a given power supply (and thus know in the case of a stack, for example, if it is worn or not). In this case, we should place the tip of the red cable to the positive terminal of the battery and the black on the negative and proceed in the same way, selecting the size and range to be measured.

To measure the resistance of a component, the component must maintain offline to not receive any current circuit. The procedure for measuring resistance is quite similar to measure voltages: simply connect each terminal of the component to meter leads (if the component is polarized, such as diodes and some capacitors, the red wire has connected to the positive terminal to the negative black component and, if the component does not has polarity, it is indifferent) and place the switch in the ohms position and scale appropriate to the size of the resistance to be measured. If you do not know about the range of resistance to be measured, we start by placing the wheel on the largest scale, and we will reduce the scale until we find the one that gives us precision without leaving range. Similarly, if the scale is chosen to be less than the value to be measured, the display shows "1" on your left; in that case, therefore, there will have to raise the range to find the right.

To measure the current flowing through a circuit, connect the multimeter in series with the circuit in question. Therefore, to measure currents have to open the circuit to sandwich the meter in the middle, so that the current circulating therein. Specifically, the process to follow is: insert the red cable into the appropriate socket (mA or A depending on the amount of current to be measured) and the black wire into the black socket, splice each wire multimeter in each of the two ends we have open circuit (closing it so therefore) and set the dial to the appropriate size and range.

Ideally, the meter operating as current meter has zero resistance to current flow through it (precisely to avoid alterations in the extent of the value of the actual intensity), so it is relatively unprotected very high currents and can easily damaged. We must always bear in mind therefore the maximum current that can support, which has both the manufacturing (besides the maximum time that may be operating in this mode).

To measure the capacitance of a capacitor, we may also use most digital multimeters market. Just we have to connect the pin capacitor to special sockets for it, marked with the brand "CX". The capacitors must be discharged before connecting to those sockets. For capacitors having polarity should be identified corresponding to each pole socket in the manufacturer's manual.

To measure continuity (ie to check whether two points of a circuit are electrically connected), simply must set the switch to the position marked with the sign of an "audio waveform" and connect the two cables to each point to be measured (polarity does not matter). Please note that this mode can only be used when the circuit under test is not receiving power. If there is continuity, the meter beeps (thanks to a buzzer that has a built); if not, you do not hear anything. You can also see the display in the display depending on the case, but the actual message depends on the model, so you should consult the instructions for each particular device.

A practical application of the multimeter function as continuity tester is to check what a breadboard holes belonging to the same node maintain its connectivity, since after continued use is relatively easy it is to break down.

WHAT IS AN ELECTRONIC SYSTEM?

An electronic system is a set of: sensors, processing and control circuitry, power supply and actuators.

The sensors receive information from the external physical world and transform it into an electrical signal that can be handled by internal control circuitry. There are all kinds of sensors: temperature, humidity, motion, sound (microphones), etc.

The internal circuitry of an electronic system processes the electrical signal conveniently. The manipulation of the signal depends on both the design of the various hardware components of the system as the logical set of instructions (ie, the "program") that the hardware be prerecorded and be able to run autonomously.

The actuators convert the electrical signal processing finished by internal circuitry in energy acts directly on the external physical world. Examples of actuators are: an engine (mechanical energy), a bulb (light energy), a loudspeaker (acoustic energy), etc.

The power supply provides the necessary power so that it can make the whole process described in "obtaining environmental information <->

Processing <-> action on the environment. " Examples of sources include batteries, AC / DC, etc.

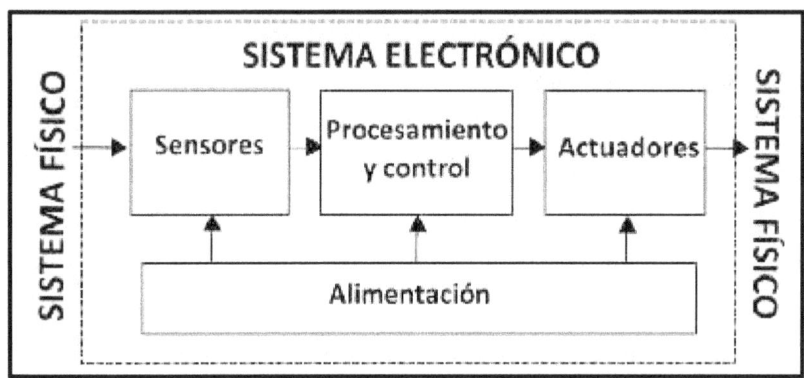

What's a Microcontroller?

A microcontroller is an integrated circuit or "chip" (i.e., an electronic device that integrates in a single encapsulated many components) having the characteristic of being programmable. That is, it is capable of running autonomously a series of instructions previously defined by us. In the above diagram representative of an electronic system, the microcontroller would be the main component of the processing and control circuitry.

By definition, a microcontroller (also commonly called "micro")
It should include within three basic elements:

CPU (Central Processing Unit) is in charge of executing each instruction and checking that the execution is successful. Typically, these instructions make use of previously available data (the "input"), and as a result generate various other data ("data output"), which may be used (or not) with the following statement.

Different types of memory: they are generally responsible for housing both the instructions and the various information they need. In this way possible that all this information (instructions and data) is always available to allow the CPU to access and work with it in

any moment. Generally we find two types of memories: that your content is stored permanently even after power cuts (called "persistent"), and that its contents are lost to stop receiving power (called "volatile"). According to the characteristics of the information to be saved, this will be recorded in one or another type of memory automatically, usually.

Different pin E / S (input / output) are responsible for communicating with the external microcontroller. In the microcontroller input pins we can connect sensors so that it can receive data from its environment, and its output pins we can connect actuators to the microcontroller can send commands and to interact with the physical environment. However, many pins on most microcontrollers are not limited input or output, but can be used interchangeably for both purposes (hence the name of E / S).

That is, a microcontroller is a complete computer (although with limited features) on a single chip, which specializes in constantly execute a predefined set of instructions. These instructions should be given at all times the information obtained and sent by device I / S and react accordingly. Logically, the instructions will vary according to the use you want to give to the microcontroller, and we decide what we are.

Increasingly household products incorporating some type of microcontroller in order to substantially increase its performance, reduce size and cost, improve reliability and reduce power consumption there. Thus, we can find microcontrollers in many electronic devices we use in our daily lives, as they can be from a simple bell to a complete robot through toys, refrigerators, televisions, washing machines, microwaves, printers, system boot of our car , etc.

WHAT IS ARDUINO?

Arduino (http://www.arduino.cc) is actually three things:

A free board hardware incorporating a reprogrammable microcontroller and a number of pin-female (which are connected internally to the
pin I / S microcontroller) for connecting there very easy and comfortable way different sensors and actuators.

When we speak of "hardware board" we are referring specifically to a PCB (for "printed circuit board", ie printed circuit board). The PCBs are made of a nonconductive material surface (normally resins reinforced fiberglass, ceramic or plastic) on which there are laminated ("glued") tracks of conductive material (usually copper). The PCBs are used to electrically connect, via the conducting paths, different electronic components soldered to it. A PCB is the most compact and stable way to build an electronic circuit (as opposed to a breadboard, perfboard or similar) but, unlike these, once made, its design is quite difficult to change. Thus, the Arduino is simply a PCB design that implements a given internal circuitry.

However, when we speak of "Arduino" we should specify the particular model, as there are several official Arduino boards, each with different characteristics (such as physical size, the number of pin-female offered, the embedded microcontroller model - and as a result, inter alia, the amount of memory utilizable-, etc.). Should know these features to identify what Arduino is that we agree more on each project.

However, although they may be different specific models (as just discussed), the microcontrollers incorporated in different Arduino boards all belong to the same "family technology" so that its operation is actually quite similar to each other. Specifically, all AVR microcontrollers are kind microcontroller architecture developed and manufactured by the brand Atmel (http://www.atmel.com). That is why, in this book we will continue appointing "Arduino" to any of them until it is essential to make some distinction.

The hardware design of the Arduino board is originally inspired by other existing hardware free board, the Wiring (http://www.wiring.co) plate. This board was created in 2003 as a personal project of Hernando Barragán, student at the time of the Design Institute Ivrea (where arose in 2005 precisely Arduino).

Software (more specifically a "development environment") free, free, cross-platform (because it runs on Linux, MacOS and Windows) that we
install on your computer and allows us to write, check and save ("load") in memory of the Arduino microcontroller instruction set that we want this start running. That is: we can program it. The standard way to connect your computer with your Arduino to send and grabarle such instructions is through a simple USB cable, thanks to most Arduino boards incorporate a connector of this type.

Arduino projects can be independent or not. In the first case, once programmed the microcontroller, the plate need not be connected to any computer and can operate autonomously if it has any power supply. In the second case, the plate must be connected in a permanent form (for USB cable, Ethernet cable, etc.) to a computer running a specific software to allow communication between it and the plate and the exchange of data between both devices. This particular software program generally what we ourselves by any standard programming language like Python, C, Java, PHP, etc., and will be completely independent Arduino development environment, which is no longer needed once the plate and It has been programmed and is operating.

A free programming language. By "programming language" means any artificial language designed to express instructions (following a certain syntax rules) that can be performed by machines. Specifically within the Arduino language, like elements are many existing programming languages (such as conditional blocks, repetitive blocks, variables, etc.) as well as different commands
-asimismo called "orders" or "functions" - that allow us to specify
in a consistent manner and without errors the exact instructions that we program into the microcontroller board. These commands write using Arduino development environment.

Both the development environment as the Arduino programming language are inspired by other existing free language and environment: Processing (http://www.processing.org), originally developed by Ben Fry and Casey Reas. The Arduino software is so similar to Processing is no coincidence, as this specializes in facilitating the generation of real-time imaging, animation and visual interactions, so many teachers Design Institute Ivrea used it in their classes. As it was in the center where it was invented precisely Arduino is natural that both environments and languages saved
quite similar. However, we should clarify that the language is built internally Processing written in Java code, while the Arduino language is based internally code C / C ++.

Arduino can perform many varied range of projects, from robotics to automation, to environmental monitoring sensors telematics, navigation systems, etc. Really, the possibilities of this platform for the development of electronic products are practically endless and are only limited by our imagination.

WHAT IS THE ORIGIN Arduino?

Arduino was born in 2005 in the Interactive Design Institute Ivrea (Italy), academic center where students were engaged in experimenting with the interaction between humans and different devices (many based on microcontrollers) for generating unique spaces, especially art. Arduino came in from the need for a device for use in classrooms that were low cost, which would work on any operating system and that counted with documentation suited to people who want to start from scratch. The original idea was thus to manufacture the plate for internal use by the school.

However, the Institute was forced to close its doors precisely
2005. Faced with the prospect of losing everything forgotten Arduino development project that had been carried out during that time, it was decided to release it and open it to the "community" that everyone had the opportunity to participate in the evolution project suggestions and propose improvements and keep "alive". And so it was: the collaboration of many people has made Arduino has gradually become what it is today: a proposed hardware and free software worldwide.

The main responsible for the idea and design of Arduino, and the visible head of the project is called "Arduino Team" formed by Massimo Banzi (teacher at the time of the Ivrea Institute), David Cuartielles (professor at the School of Arts and Communication University of Malmo, Sweden), David Mellis (then a student at Ivrea and currently member of the High-Low Tech research at the MIT Media Lab), Tom Igoe (professor at the Tisch School of Arts in New York) and Gianluca Martino (responsible manufacturer of prototype boards, whose official website is: http://www.smartprojects.it).

There is a 30-minute documentary very interesting, which is involved in the "Arduino Team" first person explaining the whole process of gestation and evolution of the Arduino project from the technical details to be taken into account to the free philosophy that permeated (e permeates) its development through different testimonies of partners worldwide. You can see for free on http://arduinothedocumentary.org.

WHAT YOU MEAN SEA ARDUINO "free software"?

In previous paragraphs we commented that Arduino is a plate of "free hardware" and "an environment and programming language (ie, software) free". But what does it mean here the word "free" exactly?

According to the Free Software Foundation (http://www.fsf.org) organization responsible for promoting the use and development of free software worldwide software has to be considered free to offer any person or organization four basic freedoms and essential:

Freedom 0: The freedom to run the program for any purpose and in any computer system.

Freedom 1: The freedom to study how the program works internally, and adapt to the particular needs. Access to the source code is a precondition for this.

Freedom 2: The freedom to distribute copies.

Freedom 3: The freedom to improve the program and release the improvements to others, so that the whole community benefits. Access to the source code is a precondition for this.

A program is free software if users have all these freedoms. Thus, free software is one software that gives users the freedom to run, copy and distribute (to anyone and anywhere), I study, change and improve it, without having to ask or pay for permission to original developer or any other specific entity. The distribution of the copies can be with or without own modifications, and attention can be free or not !: "free software" is a matter of liberty, not price.
For a program to be considered free for legal purposes has to undergo some sort of distribution license, including the GPL (General Public License) or LGPL, among others are. The issue of various licenses is a bit complicated: there are lots and lots of clauses. For more on this topic, please consult http://www.opensource.org/licenses/category, where the original official text of the major licenses available. Free software examples are many: the Linux kernel, the Firefox browser, office suite LibreOffice, VLC media player, etc.

The Arduino software is free software that is published by a combination of the GPL (for visual programming environment itself) and the LGPL (for power management and more control of the microcontroller internal codes). The consequence of this is, in short, anyone who wants (and know), may be part of the Arduino software development and thus contribute to improving the software, adding new features, suggesting ideas for new features, sharing solutions to possible existing errors, etc. This manner of operation caused the spontaneous creation of a community of people working each other over the Internet, and makes the Arduino software evolves as the community itself decides. This goes far beyond the simple question of whether the Arduino software is free or not, because the user is no longer a taxpayer to become (if you want) and an active participant in the project subject.

WHAT YOU MEAN SEA ARDUINO "free hardware"?

Free hardware (also called "open-source" or "open source") shares many of the principles and methodologies of free software. In particular, free hardware allows people to study it to understand how it works, modify, reuse, improve it and share those changes. To achieve this, the community must be able to access files on the schematic design of the hardware in question (which are CAD type files). These files detailing all the information necessary for anyone with materials, tools and knowledge to rebuild the hardware on their own without problems, since at these files could be found which individual components which make up the hardware and interconnections between each from them.

Arduino is a free hardware because its schematic files are available for download from the project website with Creative
Commons Attribution Share-Alike (http://es.creativecommons.org/licencia), which is a free license that allows derivative works both personal and commercial (provided they give credit to Arduino and publish their designs under the same license). Thus, self can build your own Arduino "by hand". However, as normal it is to buy from a dealer and pre-assembled and ready to use; in that case, of course, Arduino, albeit free, can not be free, as it is a physical object, the manufacturing costs money.

Unlike the free software world, where the ecosystem of free licenses is very rich and varied, in the field of hardware are not yet practically free hardware licenses specifically, since the concept of "free hardware" is relatively new. In fact, until recently there was a general consensus on its definition. To begin to remedy this situation, in 2010 the project OSHD (http://freedomdefined.org/OSHW), which aims to establish a collection of principles to help identify as "free hardware" a physical product emerged. OSHD is not a license (ie, legal contract), but a declaration of intent (ie, a general list of rules and characteristics) applicable to any physical device for it to be considered free. The aim of the OSHD (in whose writing has been involved people connected to the Arduino project, among others) is to provide a framework which on the one hand respect the freedom of creators to control their technology while established the adequate mechanisms for sharing knowledge and promoting trade through the open exchange of designs. In other words, show that there may be an alternative to patents for hardware that is not necessarily the public domain. The project OSHD thus opens the way to create legal precedents for facilitating the next logical step in the process: the creation of free licenses hardware.

The goal of free hardware is, therefore, facilitate and bring electronics, robotics and ultimately the current technology to the people, not a merely passive consumer, but actively involving the end user to understand and get more value from existing technology and even offering the possibility of participating in the creation of future technologies. Basically, the open hardware means having the ability to look at what's inside of things, and that's ethically correct. Allows, ultimately, improve the education of people. So the concept of free software and hardware is so important, not only for the world of computer and electronics, but for life in general.

WHY CHOOSE ARDUINO?

There are many other plates of different manufacturers, although different models incorporate microcontrollers, are comparable and offer a more or less similar to Arduino boards functionality. All of them also come with a pleasant and comfortable environment and development of a language simple and complete programming. However, the Arduino (hardware + software) platform offers a number of advantages:

Arduino is free and extensible: this means that anyone wanting to expand and improve both the hardware design of the plates as the software development environment and programming language itself, can do so without problems. This allows the existence of a rich "ecosystem" of extensions, both variants of unofficial plates as third-party software libraries that can better adapt to our specific needs.

Arduino has a large community: many people use it, and continuously enrich the documentation share their ideas.

Its multiplatform programming environment: You can install and run on Windows, Mac OS X and Linux. Not so with the software of many other boards.

Their environment and the programming language are simple and clear: they are very easy to learn and use, as well as flexible and comprehensive for advanced users can take advantage and squeeze all the possibilities for hardware. In addition, they are well documented, with detailed examples and lots of projects published in different formats.

Arduino boards are cheap: the standard Arduino (called Arduino UNO) and pre-assembled and ready for operation costs about 20 euros. Even oneself he could build (Arduino hardware is free, remember) acquiring the components separately, bringing the total price of the resulting plate would be even lower.

Arduino boards are reusable and versatile: reusable because they can leverage the same board for various projects (as it is very easy to disconnect, reconnect and reprogram), and versatile because Arduino boards provide several different types of input and output data which allow to capture information from sensors and send signals to actuators in many ways.

FEATURES MICRO Arduino UNO

It has already been discussed above there are several types of Arduino boards, each with specific characteristics that need to know to choose the model that suits us as appropriate. However, there is a "standard" model plate, which is the most widely used by far and that is what we will also use this book in all projects: Arduino UNO. Since appearing in 2010 he has undergone three revisions, so that the current model is usually called simply UNO UNO Rev3 or R3.

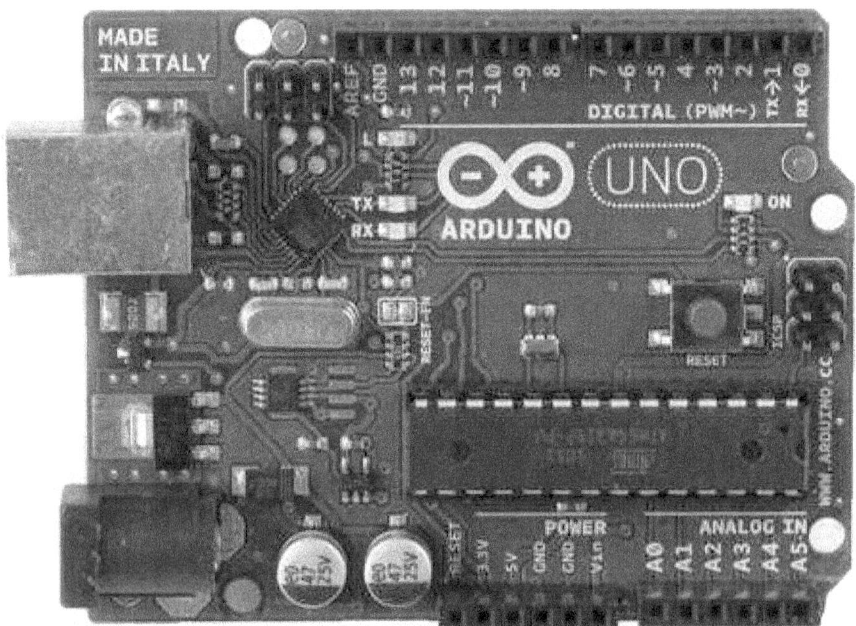

The encapsulated microcontroller

The above picture shows the Arduino UNO in its conventional variant. The following figure shows the variant called Arduino UNO SMD. The only difference between the two plates is the physical package of the embedded microcontroller: both have the same model, but the conventional plate leads mounted DIP format ("dual in-line package") and the plate SMD leads in SMD format (" Surface Mount Device "). As can be seen in both figures, the DIP format (visually, a large rectangle in the center-bottom-right of the plate) is much larger than the format SMD (visually, located in a small square in the center-diagonal bottom-right of the plate).

An important difference between the SMD and DIP format is that the former is welded to the plate surface (precisely using a technology called "surface mount", in English, SMT, "surface mount technology" -), while the second is connected to the plate through a series of metal pins (which are, in fact, device I / S microcontroller) that can be easily separated and allowing the microcontroller substitution by another if necessary. In practice, this should not matter much to us unless we want to separate and reuse our board microcontroller on other boards or assemblies; in that case, we should choose the DIP format.

Although only Arduino microcontroller communicate to us to our encapsulated these two alternatives (DIP or SMD), the world of chips in general is not so simple, because there are many variants of the two previous encapsulated, and other types of different packages. This is due to the different needs that exist regarding the availability of space and the arrangement of the connectors on the PCBs we use. If you want to see what are the most important encapsulated in the electronic world, you can download http://goo.gl/OU47S a brief document that summarizes. If you still want a more comprehensive, http://www.siliconfareast.com/ic-package-types.htm recommend consulting the page.

However, projects in this book we will not worry too much about the encapsulated chips used as working with them we will use plates "breakout". A plaque "breakout" is a leading PCB soldered a chip (or more) with the connectors and circuitry necessary to allow it to plug in external devices easily and quickly. Therefore, when we need a chip on our projects in particular, resort to any breakout board that incorporates it and then we just use the connectors offered by it to put the chip in communication with the rest of the circuit. So, just in case you need to solder a single chip to a plate is when the encapsulation type you should consider, but this is an advanced topic that will not address.

Clarify that other simple components that are not integrated circuits (such as resistors, capacitors, diodes, fuses, etc.) also may be encapsulated in SMD format to optimize the physical space occupied. In these cases, the variety of shapes and sizes, although standardized, is immense, and is beyond the scope of this book to delve into this subject. Suffice to say that

many of these encapsulations are usually distinguished by a four digit code, the first two of which indicate the length of the component and the last two in width, in hundredths of inches. For example, encapsulation (for another common part) "0603" indicate that the component in question has a size of
0.06 "x 0.03". Other common encapsulated are "0805" and "0402".

The microcontroller model

The leading microcontroller Arduino UNO is the model of Atmel ATmega328P brand. The "P" at the end means that this chip incorporates "picoPower" (owner of Atmel) technology which enables significantly lower power consumption compared to the equivalent model without "picoPower" the Atmega328 (without the "P"). However, although the ATmega328P can work at a lower voltage and consume less power than the Atmega328 (especially in Hibernation), both models are functionally identical.

As is the case with the rest of microcontrollers used in other Arduino, the ATmega328P has a type AVR architecture, architecture developed by Atmel and to some extent "competing" for other architectures such as Microchip PIC manufacturer. More specifically, it belongs to the subfamily ATmega328P microcontroller "megaAVR". Other subfamilies of the AVR architecture are the "tinyAVR" (whose microcontrollers are more limited and are identified with the name of ATtiny) and "XMEGA" (whose microcontrollers are better able to identify and name ATxmega), but not the study as the Arduino microcontrollers plates do not incorporate these families.
Anyway, if you want to know more about the AVR architecture and models and features offered microcontrollers built this way, there is nothing better than to consult the website of the manufacturer itself: http://www.atmel.com /products/microcontrollers/avr/default.aspx). And more specifically, if you want to read the technical specification ATmega328P (although for the projects in this book will not be necessary) can be downloaded here: http://www.atmel.com/dyn/resources/prod_documents/doc8161.pdf .

So if we can come in handy it is to know the specific pinout (also called "pin") Input / output of the microcontroller, because although we have said before that in general all the pins of E / S serve to communicate the microcontroller with the outside world, it is true that each pin tends to have a certain specific function. As each model of microcontroller has a different number and location of pins, in our case we have at hand the pin arrangement ATmega328P. The following figure shows this provision in the DIP encapsulation type, and has been obtained from the technical specification mentioned in the previous paragraph. Note: the circle that appears at the top of the figure indicates where there is a notch in the actual encapsulation, so that it easy to distinguish the orientation of the pins.

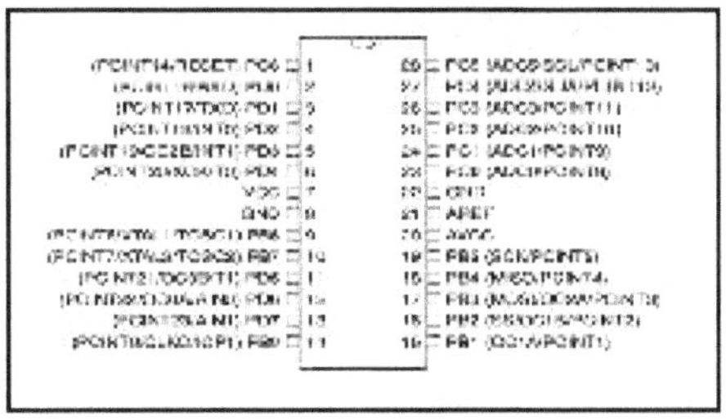

Noting the image can know which pin is receiving power (designated as "VCC"), which two pins are connected to ground (those listed as "GND"), which pins are I / S (denoted as PBx, PCx or PDx) and the existence of other more specific as the AVCC pin (where supplementary feeding for internal analog-digital converter chip) or the AREF (analog reference which is received for that -this converter is received it study further on). You may also notice that alongside the name of the pin I / S is indicated in parentheses specialized functions that each of them has in particular (in addition to its generic function input / output). Some of these we will study the specific functions throughout the book, such as the role of "reset" the microcontroller, or communication with the outside using the serial protocol or SPI or I2C, or using interrupts, or the outputs of the PWM, etc.

The memories of the microcontroller

Another thing to be aware of microcontrollers are the types and amounts of memory housed inside. In the case of ATmega328P we have:

Flash memory: persistent memory where the program that runs the microcontroller (up to a new rewrite if need be) stored permanently. In the case of ATmega328P has a capacity of 32KB.

In microcontrollers that are included in the Arduino we can not use the full capacity of the Flash memory because there are 512 bytes (called "bootloader block") already occupied by a factory preset code (called "bootloader" or "manager Boot "), which allows us to use the Arduino board in a simple and convenient way without having to know the most advanced electronic internals of the microcontroller. The ATmega328P that can be purchased individually factory normally exclude this small program, so do offer 32KB integrity, but we can not change

wait connect to an Arduino board and functioning as lacking without having them recorded that "preset". From boot loaders, its use and its importance in the next section we'll talk.

SRAM memory: volatile memory where at that moment the program (recorded separately in the Flash memory, remember) need to create or manipulate for correct operation are housed. These data are often variable content throughout the runtime of the program and each is of a specific type (ie, a data can contain an integer number value, another a decimal number, one a character value ... also can be fixed text strings or other more special data). Whatever type of data, their value will always be removed when you stop supplying power to the microcontroller. In the case of ATmega328P this memory has a capacity of 2KB.

If we need to extend the amount of available SRAM, we could always acquire independent SRAM and connect to the microcontroller using a communication protocol known for this (as SPI or I2C, which we'll talk soon); however, this will not be necessary in the projects in this book.

EEPROM memory: persistent memory where data to be recorded once staying off the microcontroller so that they can then use on these reboots are stored. For the ATmega328P this memory has a capacity of 1 KB, so it can be understood as a table of 1024-byte positions each.

If we need to expand the number of available EEPROM memory, we can always acquire independent EEPROM and connect to the microcontroller using a communication protocol known for this (as SPI or I2C, which we'll talk soon). Or, alternatively, purchase memory cards of type SD ("Secure Digital") and communicate via a dedicated circuit to the microcontroller. SD memories are actually simple Flash memories, encapsulated in a particular way; They are widely used in digital still cameras / video and next-generation mobile phones as they offer lot of capacity (several gigabytes) for a cheap price. The reason why these cards are capable of being recognized by the ATmega328P is because they can operate using the SPI communication protocol.

We can deduce from the above that the architecture belongs to the chip ATmega328P (and in general, the whole family of microcontrollers AVR) is Harvard type. In this architecture, the memory hosting the data (in our case, SRAM or EEPROM) is separated from the memory that holds the instructions (in our case, Flash), so both memories communicate with the CPU completely independently and in parallel, thus achieving greater speed and optimization. Another type of architecture (which is what we see in PCs) is the Von Neumann architecture, in which the CPU is connected to a single RAM containing both program instructions and data, so the operation speed is limited (among other things) by the bottleneck effect it means a single communication channel for data and instructions.

The microcontroller records

Records are available memory space within the local CPU of the microcontroller. They are very important because they have several essential functions: they serve to hold the data (previously loaded from the SRAM or EEPROM memory) required for the execution of instructions and forthcoming (and so perfectly have them available at the right time); also they serve to temporarily store the results of the recently executed instructions (if needed at some later time) and also serve to accommodate themselves to the instructions on the spot are running.

Its size is very small: only have the capacity to store a few bits each. But this factor is one of the most important features of any microcontroller, since the larger the number of bits "fit" in their records, the better its performance in terms of computing power and speed of execution. Indeed, it is easy to see (to oversimplify) that a microcontroller with records twice as large as another can process twice as much data and therefore work "twice as fast" even running both at the same pace. In fact, this feature is so important that when we hear that a microcontroller is "8-bit" or "32-bit", we are referring precisely to this fact: the size of your records.

Depending on the value that we give to the microcontroller, it is necessary to use one with a size sufficient records. For example, control of a simple appliance such as a blender requires no more than a microcontroller 4 or 8 bits. Instead, the electronic control system car engine or the ABS brake system are typically based on a microcontroller 16 or 32 bits.

The ATmega328P chip is 8 bits. In fact, all microcontrollers that incorporate different Arduino boards are 8 bits except the built-in Arduino Due, which is 32 bits.

Communication protocols I2C / TWI and SPI

When you want to transmit a set of data from an electronic component to another it can be done in many ways. One is establishing communication "series"; in this type of communication information is transmitted bitwise (one after another) through a single channel, thereby sending one bit at a time. Another way to transfer data is by calling communication "parallel", in which several bits are sent simultaneously, each on a separate channel and synchronized with the rest.

The microcontroller, through some of its pin I / S, uses the serial communication system to transmit and receive commands and data to / from other electronic components. This is due mainly to a series communication in theory need only a single channel (one "cable"), while in a number of parallel communication cables are needed, with a corresponding increase in complexity, size and cost resulting circuit.

However, we can not speak of a single type of serial communication. There are many different protocols and standards are all based on the serial data transfer, but implemented differently each specific details (such as the mode of synchronization between transmitter and receiver, the transmission speed, the size of the packages data, connection and disconnection messages and give way to another in the exchange of information, the voltages used, etc.). Among the large number of serial communication protocols recognized by the vast variety of electronic devices on the market, which we know are interested that the ATmega328P is able to understand and therefore which can be used to contact the variety of peripherals . In this regard, the most important standards are:

I2C (Inter-Integrated Circuit, also known as TWI-of "two-wire", literally "two cables" in English) is a system widely used in industry mainly integrated circuits to communicate with each other. Its main feature is that it uses two lines to transmit Information: one (called line "SDA") is used to transfer data (0s and 1s) and another (called line "SCL") used to send the clock signal. Actually two lines are also needed: the power and common ground, but these presuppose existing in the circuit.

By "clock" a very precise binary signal of a periodic frequency used to coordinate and synchronize the constituent elements of a communication (ie, transmitters and receivers) so that everyone knows when it starts, how long it lasts and when means just information transfer. In data sheets and diagrams to the clock signal in general it is often described as CLK (for "clock").

Each device connected to the I²C bus has a unique address that identifies about other devices, and can be configured as "master" or "slave". A master device is the one that initiates the transmission of data and also generates the clock signal, but it is not necessary that the master device is always the same: this feature can be neatly exchanging the devices with that capability.

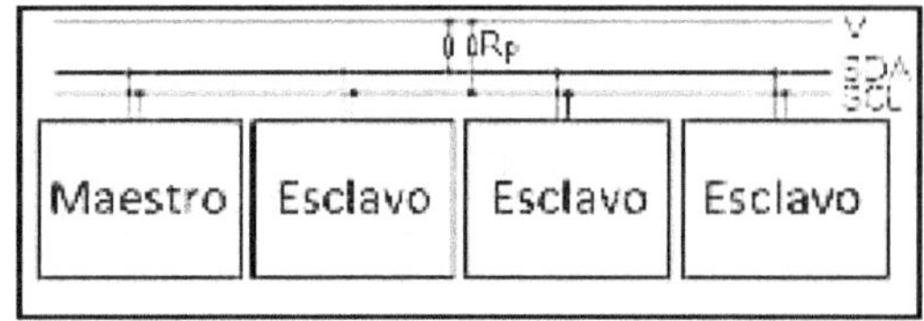

As shown in the diagram above, to function properly both the "SDA" line as the "SCL" need to be connected via a resistor "pull-up" to the common power supply, which can provide a voltage generally 5 V or 3.3 V (although systems with other voltages are possible).

The data transfer rate is 100 Kbits per second in standard mode (though also allow speeds up to 3.4 Mbit / s). However, having a single data line, the transmission of information is "half duplex" (ie, communication can only be set in the same direction time) so that when a device starts receiving a message, will have to wait for the transmitter to stop transmitting to respond.

SPI (Serial Peripheral Interface): As the I²C system, the communication system SPI is a standard that allows control (short distances) almost any digital electronic device that accepts a bit stream synchronized series (that is to say, regulated by a clock). Similarly, a device connected to the SPI bus can be "master" -in English, "master" - or "slave" -in English, "slave" - where the first is the one that initiates the transmission of data and also generates the clock signal (although, as with I²C, SPI is also necessary that the teacher is always the same device) and second only it responds.

The biggest difference between SPI and I²C protocol is that the former requires four lines ("wires") instead of two. A line (usually called "SCK") sends the clock signal generated by the current master of all devices; another (usually called "SS") is used by the teacher to choose at all times how slave device you want to communicate among several that may be connected (since it can only transfer data with one slave at a time); another (usually called "MOSI") is the line used for sending data -0s And 1S- from the master to the slave chosen; and the other (usually called "MISO") is used to send data to the contrary: the response of the slave to the master. It is easy to see that having two lines for data transmission information is "full duplex" (ie, the information can be transported in both directions simultaneously).

The following figures show the scheme of lines of communication between a master and a slave and between a master and three slaves respectively shown. It can be seen that, in the case of the existence of several slaves is necessary to use a different line "SS" for each of them, since this line is used to activate the particular slave that at all times the master you want to use (this does not happen with the clock lines, "MOSI" and "MISO" that are shared by all devices) .Técnicamente speaking, the slave who receives for its line SS a value lower the voltage that is selected at that time by the teacher, and those who receive the highest value will not (hence the higher underline that appears in the figure).

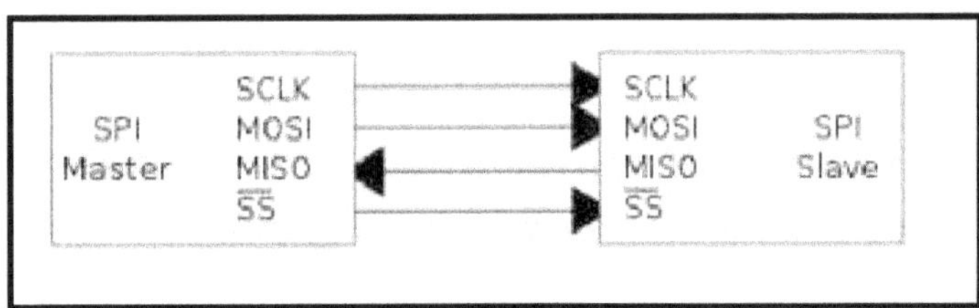

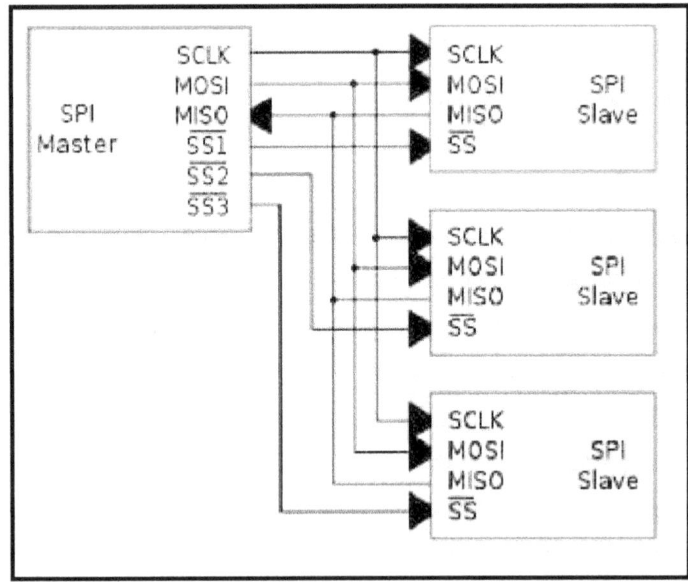

As you can see, the SPI protocol for the I^2C has the disadvantage of requiring the microcontroller pin devote many more I / S to external communication. Instead, we highlight the advantage that it is faster and consumes less power than I^2C.

As can be seen from the figure showing the arrangement of ATmega328P microcontroller pins (page 75), the pins corresponding to the lines SDA and SCL I2C are the numbers 27 and 28, respectively, and pins corresponding to the SPI lines SS , MOSI, MISO and SCK are numbers 16, 17, 18 and
19, respectively. If more lines SS are needed (because you have more than one slave device connected to our circuit), one could use any pin I / S that he respects the agreement put the value of its output voltage LOW when desired work associated with the slave device and put the rest ALTO SS pin.

The bootloader of the microcontroller

We have already mentioned that in the Flash memory of the microcontroller included in the Arduino board comes pre factory a little program called "bootloader" or "boot loader" which is essential for comfortable and easy handling of the plate in question. This software (also called "firmware" because it is a type of software that rarely changes) occupies, in the Arduino UNO, 512 bytes of space in a special section of Flash memory, called "bootloader block", but other models can occupy more Arduino boards (eg in the Leonardo model occupies 4
Kilobytes).

The function of this firmware is automatically manage the process of recording in the Flash memory of the program that we want the microcontroller run. Logically, the bootloader made this recording beyond the "bootloader block" not overwritten himself.

More specifically, the bootloader is responsible for receiving part of our program Arduino development environment (usually via a transmission via USB connection from the computer that is running this environment up to the plate) to then proceed to its proper storage Flash, all memory automatically and without us having to worry us of electronic insides of the process. Once completed the recording process, the bootloader terminates and the microcontroller has processed immediately and permanently newly recorded instructions (with power on).

In the Arduino UNO, the bootloader always executed during the first second of each restart. During these moments, the boot loader is expected to receive a number of specific instructions from the development environment to interpret and make the corresponding burden of a possible program. If these instructions do not arrive after that time, the bootloader finishes its execution and also begin to process what you have at that time in the Flash memory.

These internal instructions for programming of flash memories are slightly different depending on the type of bootloader that has the plate, but almost all are variants of the instruction set officially offered by Atmel for programming their microcontrollers, called STK500 protocol (http: // www.atmel.com/tools/STK500.aspx). One example is the bootloader that has prerecorded ATmega328P Arduino UNO, based on a free firmware called Optiboot (http://code.google.com/p/optiboot), which achieves a recording speed of 115 kilobits per second program to be loaded through the use of instructions derived own "standard" STK500. Another example of derivative bootloader is the bootloader STK500 protocol "wiring" recorded factory microcontroller Arduino Mega. The bootloader that comes in the Leonardo plaque (called "Catherine") is different because it meant another independent set of instructions called AVR109 (also officer Atmel). All this information can be obtained by viewing the contents of the file named "boards.txt" downloaded with the Arduino development environment.

If we acquire ATmega328P separate microcontroller, keep in mind that not have the bootloader, so we incorporate one ourselves "in place" to make use of it thereafter, or never use any bootloader and load time always our programs to Flash memory directly. In both cases, the procedure requires the use of a specific device (specifically, what is called a System -In Programmer- "ISP programmer") that must be purchased separately. This device must be connected on one side to our computer and the other to the Arduino, and supplies the absence of bootloader making intermediary between our development environment Flash memory and microcontroller. Therefore, we can summarize by saying that the bootloader is the element that enables programming our Arduino directly with a simple USB cable and nothing else.

For convenience, in the installer package Arduino development environment (downloadable from its official website, for more details see the next chapter), bit-exact official bootloaders that are recorded in different Arduino microcontrollers are also distributed copies. These files are exact copies of "hex" extension having an internal format called "Intel Hex Format". For normal use of our board we do not need at all these files "hex" but if we had an ISP programmer and sometime we had to "replace" a bootloader damaged (or burn a bootloader to a microcontroller that had none), Arduino offers these files to load them into the Flash memory of our microcontroller whenever we want.

Intel Hex Format The format is used for all AVR chips to store their content in Flash memory, so it should be clarified that not only the bootloaders are housed internally in this way Flash memory, but our own programs write in the development environment will also be housed there in "hex" format (although these details should not we worry about now).

Clearly, Arduino bootloaders are also free software, so as occurs with Arduino programming environment, always have available the source code (written in C language) in order to know how it works internally for edltlng even if they lt deemed appropriate.

WHAT HAVE OTHER FEATURES Arduino UNO?

Arduino UNO microcontroller incorporating apart, has other interesting features to review:

Feeding

The operating voltage of the Arduino board (including the microcontroller and other components) is 5 V. we get this power in several ways:

Arduino connecting to an external source such as an AC / DC adapter or a battery. For the first case, the plate has a socket which can be plugged a plug of 2.1 mm type "jack". For the second, the outgoing cables to the battery terminals can be connected to pins-female marked "Vin" and "GND" (positive and negative respectively) in the area of the plate marked with the label "POWER". In both cases, the board is it prepared theoretically for a supply of 6-20 volts, but, really, the recommended input voltage range (taking into account the desire for a certain stability and electrical safety in our circuits) is lower: 7 to 12 volts. In any case, the input voltage provided by the external source is always lowered to working 5 V through a voltage regulator circuit that is already built into the plate.

Connecting Arduino to our computer via a USB cable. To this end, the plate has a female USB connector type B. The power thus received is permanently regulated to
5 V work and provides a maximum of up to 500 mA of current (therefore, the power consumed by the plate is in that case about 2.5 W). If at any time the USB connector spends more desirable intensity, the Arduino board is protected by a resettable Polyfuse that automatically breaks the electrical connection until conditions return to normal. A consequence of this protection against current peaks is possible that the intensity of current received through ARDUINO. Training Workshop

USB may not be enough for projects containing components such as motors, solenoids or arrays of LEDs, which consume a lot of power.

Whatever the way chosen to power the board, this is the "smart" enough to automatically select at all times the available power source and use one or the other without us having to do anything special about it.

If we use a battery as external power, the ideal would be to 9 V (is within the recommended range 7 to 12 volts), and if an AC / DC adapter is used, the use of one is recommended with the following characteristics:

The output voltage must be offered from 9 to 12 V DC. In fact, the regulator circuit that incorporates Arduino can handle output voltages (input board) up to 20 V, so in theory could be used AC / DC to generate an output of 20 V DC . However, this is not a good idea because most of the voltage as heat (which is terribly inefficient) is lost and can also lead to overheating of the controller, and consequently damage the board.

Available power intensity must be 250 mA (or more). If we connect to our Arduino many components or a few but high energy consumers (such as an LED array, an SD card or engine) adapter should provide at least 500 mA or 1 A. In this way we will ensure we have enough power to each component to function reliably.

The adapter must be polarity "with center positive." This means that the outside of the metal cylinder that forms the plug
5.5 / 2.1 mm adapter must be the negative terminal and the hollow interior of the cylinder it is to be the positive terminal. The easiest thing to make sure our adapter is right in this sense is to see if somewhere has printed the following symbol

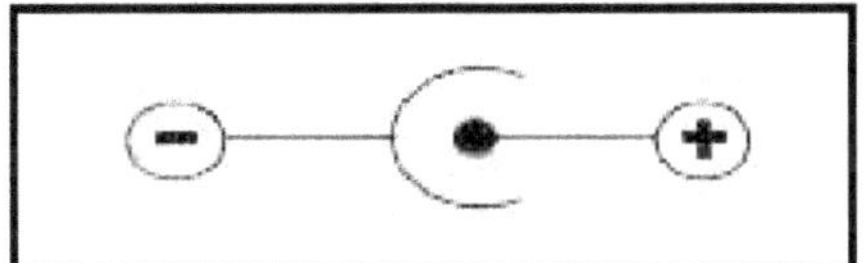

On the other hand, within the area labeled "POWER" on the Arduino board is a series of pin-female-related power, such as:

"GND" pin-female grounded. It is very important that all components of our circuits share a common ground reference. These pin-female offered to perform this function.

"Vin": this pin-female can be used for two different things: if the board is connected via the 2.1mm jack to an external source to provide a voltage within safety margins, can connect to this pin-female any electronic components for feeding directly the voltage level that is providing the power at that time (unregulated by the plate!). If the board is powered via USB, then that pin-female provide 5 V regulated. In any case, the maximum current of 40 mA is provided (this must be taken into account when connecting devices that draw a lot of current, such as engines). We can also use the pin-female "Vin" for something else to feed the board itself directly from any external power source without using plug or USB cable. This one 9V is made by connecting the positive terminal of the source (eg,
V) and connecting the negative terminal to the ground pin. If this arrangement is used, the voltage regulator that incorporates the plate will reduce the voltage received from the battery to the voltage of the plate (5 V).

"5 V": this pin-female can be used for two different things: whether the board is powered via the USB cable as it is fed by an external source to provide a voltage within safety margins, we can be connected to this pin-female any component so that it can receive 5 V regulated. In any case, the maximum intensity of generated current will be 40 mA. But we can also use the female painted for something else to feed the plate itself from an external power supply pre-set to 5 V without using the USB cable and 2.1mm plug.

"3.3 V" pin-female This provides a voltage of 3.3 volts. This voltage is obtained from either received through the USB cable or plug of 2.1 mm, and is regulated (with a margin of error of 1%) for a specific circuit incorporated in the plate: the LP2985. In this particular case, the maximum current of 50 mA is generated. As with the previous pin, we can use this pin to feed our components said voltage circuits requiring (the finest), but instead, can not connect any external source here because the voltage is too limited to feed the plate.

The chip ATmega16U2

The USB connection on the Arduino board, in addition to serving as power, especially as a means to transmit data between our computer and the board, and vice versa. This traffic information is performed between the two devices is accomplished through the use of USB protocol, a standard protocol such that both our computer and Arduino are able to understand and handle. However, the USB protocol is internally too complex for ATmega328P microcontroller can understand itself without help, since he can only communicate with the outside through technically much simpler protocols such as I2C or SPI, and few others. It is therefore necessary that the board provided a "translator" to provide the ATmega328P element (specifically, the receiver / transmitter UART TTL-series type that incorporates) the handling of the information transferred via USB without the need to know the intricacies of the protocol.

The Arduino UNO R3 board has a chip that performs this function of "translator" of the USB protocol to a simpler (and vice versa) serial protocol. That chip is ATmega16U2. The microcontroller is an all ATmega16U2 itself (with its own CPU, memory -has its own eg 16 Kilobytes of

Flash memory for internal use, hence its name, etc.) and therefore could perform many more tasks that not only the "translation" of the USB protocol. In fact un-assign it is technically possible to do other things and thus make the Arduino in virtually any type of USB device connected to our computer (keyboard, mouse, MIDI device ...). However, the default one included in ATmega16U2 Arduino firmware is already preprogrammed to perform exclusively the function of "interpreter" to ATmega328P and go.

This firmware is free software, so you can access its source code and its corresponding file "hex" are also available within the set of downloaded files, along with the environment official Arduino (specifically, within the folder "firmware" within "hardware / arduino").

In previous models Arduino UNO (such as NG model, Diecimila or Duemilanove) the ATmega16U2 chip did not come; instead featured a USB-to-serial converter manufacturer FTDI, the FT232RL circuit specifically. One advantage of having replaced the FT232RL the ATmega16U2 is price. Another advantage is having the ability (if we are advanced users, as already mentioned) to reprogram the ATmega16U2 so that instead of functioning as a simple converter USB-Series can simulate any type of USB device, for example with the FT232RL we can not because it is designed to do just the function for which it was built.

The ATmega16U2 comes with the Arduino for crystal clock oscillator, exclusively for him, used to maintain synchronization with the USB communication. Watches discussed in a later section.

The digital inputs and outputs

Arduino has 14-pin socket inputs or outputs (depending on what suits) Digital, numbered from 0 to 13. This is where we connect our sensors so that the plate can receive data from the environment, and which connect the actuators so that the plate can send the necessary orders, besides being able to connect any other components that need to communicate with the board somehow. Sometimes these digital pin-female "general purpose" GPIO pins are called (for "General Purpose Input / Output").
These pin-female 5 V digital work can provide or receive a maximum of 40 mA and have a resistance "pull-up" internal between 20 and 50 KQ KQ which is initially off (unless we indicate otherwise by programming software).

Keep in mind, however, that although each individual pin can provide up to 40 mA maximum, actually, the plate internally grouped digital pins so that they can only provide 100 mA while all the pins No. 0,1,2,3 and 4, and 100 mA over the rest of the pins (5 to 13). This means that as much could have 10 pins offering 20 mA at once.

The analog inputs

Arduino has 6 analogue inputs (in the form of female-PIN labeled "A0", "A1" ... to "A5") which can receive voltages within a range of continuous values between 0 and 5 V. However, electronics plate can only work with digital values, so a prior conversion of the analog value received as close a possible digital value is required. This is done by an analog / digital converter circuit incorporated in the plate itself.

The converter circuit is 6 channels (one for each input) and each channel has 10 bits (called "bit resolution") to save the voltage value converted digitally.

Overall, the number of bits of resolution that has a specific analog / digital converter is what makes largely the degree of accuracy achieved in the conversion from analog to digital, since the more bits of resolution have, the more accurate will be transformation. For example, in the case of built-in Arduino converter, if you count the number of combinations of 0s and 1s that can be obtained with 10 positions, we see up to 210 (1024) possible different values. Therefore, the Arduino can distinguish for the digital voltage from the value 0 to the value 1023. If the converter has eg 20 bit resolution, the range of digital values that could distinguish would be much larger (220 = 1048576) and You could refine the accuracy more.

This is easily seen if we divide the analog input range (5 V - 0 V = 5 V) between the maximum number of digital values (1024). We obtain each digital value corresponds to an analog "window" of about 5
V / 1024≈ 5 mV. In other words, all analog values within each range of 5 mV (from 0 to 5 V) is "collapse" without distinction in a unique digital value (from 0 to 1023). So, we can not distinguish analog values spaced by less than 5 mV.

In many of our projects us is enough this degree of precision, but others may not. If the analog / digital converter has more bits of resolution, the result of rango_analógico_entrada / número_ valores_digitales division would be lower, and therefore the conversion would be more rigorous. But you can not increase the bit resolution converter board, if we want more accuracy is to opt for another solution: instead of increasing the denominator of the previous division can reduce the numerator (ie, the range analog input, or more specifically, its upper limit -for equal to 5 V defect, since the lower is 0). This upper limit on official documentation is often named as "reference voltage".

The practical and concrete way to reduce the reference voltage of the analog / digital converter Arduino we can not explain yet because we still lack the necessary skills to carry out the whole process. It will be explained in the section of Chapter 6.

Finally, say that these pin-female analog input also have all the functionality of the input-output pin digital. That is, that if you ever need more digital pin-female beyond 14 Arduino offers (0 to 13), 6-pin-female analog can be used as a pin-female digital plus (numerándose then from 14 to 19) without distinction.

The analog outputs (PWM)

In our projects often need to send analog signals to the environment, for example, to vary progressively the speed of a motor, the frequency of a sound emitted by a buzzer or the intensity with which sports an LED. Not just simple digital signals: we have to generate signals that vary continuously. Arduino does not have these pin-female analog output itself (because its internal electronic system is not capable of handling this type of signals), but uses some pins female-specific digital output to "simulate" an analog behavior. Digital pin-female who are able to work in this mode are not all are only marked with the "PWM" label. Specifically for the Arduino UNO model are pin number: 3, 5, 6, 9, 10 and 11.

The acronym comes from PWM "Pulse Width Modulation" (Pulse Width Modulation). What makes this type of signal is issued, instead of a continuous signal, a square signal formed by pulses of constant frequency (approximately 490 Hz). The grace is that by varying the duration of these pulses will vary proportionally resulting average voltage. That is: the shorter the pulses (and therefore farther apart in time, and its frequency is constant), the lower the average output voltage, and the longer the pulses (and many, closer together in time are), the higher the voltage. The extreme case would have when the pulse duration coincides with the period of the signal, at which indeed there would be no distance between pulse and pulse (it would signal a constant value) and the average output voltage is the maximum possible which is 5 V. The pulse length can change at any time while the signal is being broadcast, so as a result the average voltage may shift over time continuously. These include illustrate what just explained:

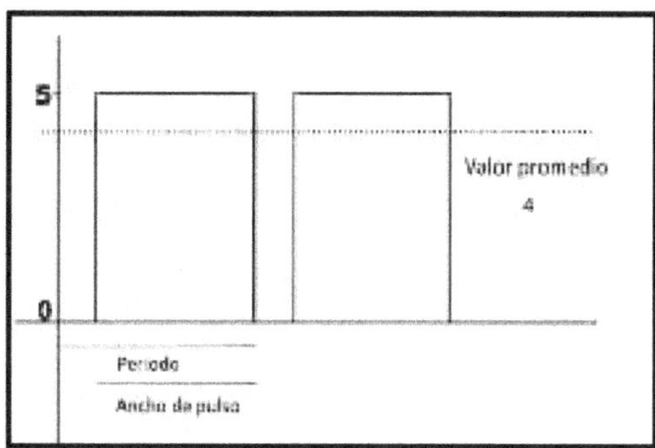

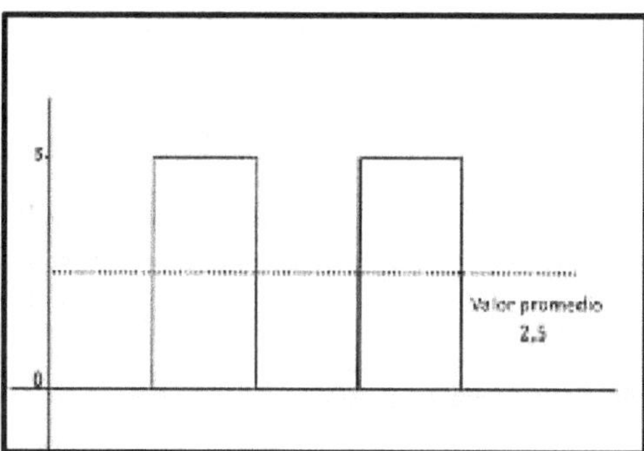

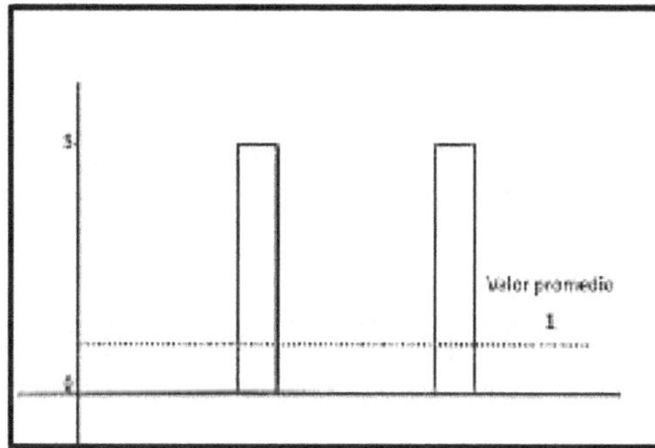

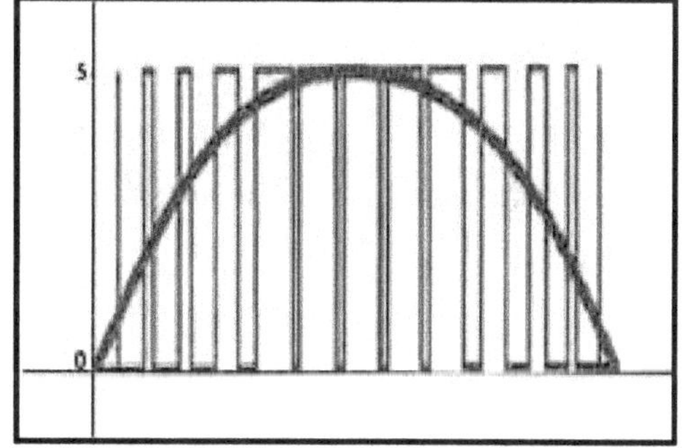

Each pin-female PWM has a resolution of 8 bits. This means that if you count the number of combinations of 0s and 1s that can be obtained with 8 positions, we get a maximum of 28 (256) different possible values. By

92

Therefore, we can have 256 different values to indicate the desired duration of the pulses of the pulse signal (or put another way: 256 different mean values). If we set (by software programming) the minimum value (0), we will be issuing some extremely narrow pulses and generate a corresponding analog signal to 0 V; if we set the maximum value (255) will emit pulses and generate a maximum equivalent to 5V analog signal therebetween any intermediate value will pulse duration and therefore generate an analog signal to a value between 0 V and 5 V .

The existing analog voltage difference between two adjacent average values (ie, including for example the value number 123 and number 124) can be calculated by dividing: rango_voltaje_salida / número_valores_promedio. In our case, it would be (5 V - 0 V) / 256 mV ≈ 19,5. That is, each average value is above and spaced from the next by a "hop" of 19.5 mV.

You can change the default frequency of the square signal used in the generation of the "analog" signal, but it is not a trivial procedure, and in most cases it will not be necessary. Here, we discuss only the PWM pin is controlled by three different timers which maintain the constant frequency of the pulses emitted; specifically, pins 3 and 11 are controlled by the "TMR1", pins 5 and 6 by the "Timer2" and the pins 9 and 10 by the "timer3".

Other uses of the female pin-plate

There are certain pin-socket digital input / output, in addition to its "standard" function, have other specialized functions. For instance:

Pin 0 (RX) and pin 1 (TX) allows the microcontroller ATmega328P can directly receive serial data (for the pin RX) or transmission (for the pin TX) without going through the USB-serial conversion performed by the chip ATmega16U2 . That is, these pins enable communication without intermediaries of external devices with the receiver / transmitter serial (TTL-UART) that incorporates ATmega328P own. However, these pins are internally connected (via resistors 1 k) to ATmega16U2 chip, so that the available data in the USB also be in these pins.

We must clarify that on the plate are embedded a pair of LEDs labeled "RX" and "TX" but that, despite their name, do not light when receive or transmit data pins 0 and 1, but only when they receive or transmit data from the USB connection through the chip ATmega16U2.

Pins 2 and 3: can be used, with the help of scheduling software, to manage interruptions. However, this issue is relatively advanced and do not discuss in this book.

Pins 10 (SS), 11 (MOSI), 12 (MISO) and 13 (SCK) can be used to connect any device with which you want to carry out communications through the SPI protocol. Later we study specific cases.

Pin 13: This pin is connected directly to an LED embedded in the plate (identified with the label "L") so that if the voltage value received by this pin is high (HIGH), the LED will light, and if the value is low (LOW), the LED will turn off. It is a simple and fast way to detect signs of external inputs without need for any extra components.

There are also a pair of pin-female analog input with an extra function besides the usual:

A4 pin (SDA) and A5 (SCL) can be used to connect any device that you want to perform communication via I2C / TWI protocol. Arduino provides (for a simple matter of convenience and ergonomics) a doubling of these two pin-female in the last two pin-pin female after "AREF", which are unlabeled because there is no physical space.

Finally, along the plate there are different pin-female not commented yet they do not work or as outputs or as inputs because they have a very specific and particular use:

AREF pin-offers an external reference voltage to increase the accuracy of the analog inputs. We study its practical use in Chapter 6.

Pin RESET: if the voltage of this pin is set to low (LOW) value, the microcontroller will reset and will start the bootloader. To perform this same function, the Arduino already has a button, but the button offers the possibility of adding another reset button plates extra (ie, connecting plates above the Arduino board to expand and complement), which by their placement can hide or lock the button on the Arduino board.

IOREF Pin: This pin is actually a regulated duplication pin "Vin". Its function is to indicate to the extra plates connected to our Arduino working voltage to the input pins / out of this, for that extra plates are automatically adapted to the working voltage (in the case of the model we know ONE which is 5 V).
Unused pin: pin just below the IOREF, which is unlabeled, currently not used for anything, but is reserved for future use.

Once known all-female pin Arduino UNO is very interesting to see what correspondence between each pin and ATmega328P microcontroller. Because in reality, most of these female-PIN you do is simply provide an easy and convenient direct connection to these pins, and little else. This can be seen in the following figure; it shows what the mapping pins Arduino regarding ATmega328P microcontroller pins.

Arduino function					Arduino function
reset	(PCINT14/RESET) PC6	1	28	PC5 (ADC5/SCL/PCINT13)	analog input 5
digital pin 0 (RX)	(PCINT16/RXD) PD0	2	27	PC4 (ADC4/SDA/PCINT12)	analog input 4
digital pin 1 (TX)	(PCINT17/TXD) PD1	3	26	PC3 (ADC3/PCINT11)	analog input 3
digital pin 2	(PCINT18/INT0) PD2	4	25	PC2 (ADC2/PCINT10)	analog input 2
digital pin 3 (PWM)	(PCINT19/OC2B/INT1) PD3	5	24	PC1 (ADC1/PCINT9)	analog input 1
digital pin 4	(PCINT20/XCK/T0) PD4	6	23	PC0 (ADC0/PCINT8)	analog input 0
VCC	VCC	7	22	GND	GND
GND	GND	8	21	AREF	analog reference
crystal	(PCINT6/XTAL1/TOSC1) PB6	9	20	AVCC	VCC
crystal	(PCINT7/XTAL2/TOSC2) PB7	10	19	PB5 (SCK/PCINT5)	digital pin 13
digital pin 5 (PWM)	(PCINT21/OC0B/T1) PD5	11	18	PB4 (MISO/PCINT4)	digital pin 12
digital pin 6 (PWM)	(PCINT22/OC0A/AIN0) PD6	12	17	PB3 (MOSI/OC2A/PCINT3)	digital pin 11 (PWM)
digital pin 7	(PCINT23/AIN1) PD7	13	16	PB2 (SS/OC1B/PCINT2)	digital pin 10 (PWM)
digital pin 8	(PCINT0/CLKO/ICP1) PB0	14	15	PB1 (OC1A/PCINT1)	digital pin 9 (PWM)

Digital Pins 11, 12 & 13 are used by the ICSP header for MISO, MOSI, SCK connections

The ICSP connector

The ICSP acronym (meaning "In Circuit Serial Programming") refers to a method for programming AVR microcontrollers, PIC and Parallax Propeller type that does not have preinstalled bootloader directly. We know that the function of a bootloader is to allow our programs to the microcontroller load plate connecting to our computer through a simple USB cable standard, but if that microcontroller has not recorded any bootloader, memory writing can not be performed in this way so simple and we must use other methods, such as ICSP.

This situation can find us when we want to replace such a DIP microcontroller Arduino UNO on the other we have acquired separately without bootloader built plate. In this case we could opt to use the ICSP method grabarle a bootloader (so that it can be re-programmed via USB directly, in fact, that's how it has been recorded precisely the bootloaders in the microcontroller Arduino boards that come standard) or also to record our programs always directly without having to use any bootloader ever (with the advantage then more free space in the flash memory microcontroller -the occupy the bootloader if estuviera- and to run our programs immediately after the plate is under power without waiting for the implementation of a nonexistent bootloader). Another situation in which we may want to use the ICSP method is when you want to overwrite the existing bootloader on the other, because the original has been corrupted, for example.

To program a microcontroller attached to any Arduino by ICSP method a specific hardware device, the "Programmer ISP" is needed. There are several forms and models, but today the most widespread version of ISP programmer is a device consisting of one side of a USB connector to plug into our computer, on the other a pin ICSP list to fit the ICSP connector our Arduino, and whose inner "heart" is a particular specialized function of microcontroller programmer.

Although everyone has a similar extensive look, keep in mind that not all ISP programmers are compatible with all models of AVR microcontrollers. Among those who are with ATmega328P microcontroller (the full list is available in the file named "programmers.txt" downloaded with the Arduino development environment) we find the following:

"AVRISP mkII" (http://www.atmel.com/tools/MATUREAVRISP.aspx) is the official manufactured by Atmel ISP programmer, strictly based on the official protocol transfer Atmel STK500. Supports virtually all models of AVR microcontrollers and development environments AVR chips most common (besides the Arduino environment). There is already above (obsolete) version of this programmer called "AVR ISP".

"USBtinyISP" (http://www.ladyada.net/make/usbtinyisp): ISP programmer manufactured by Adafruit, which aims to improve the above. For example, an advantage that has compared the "AVRISP mkII" is that, unlike this, can feed 5 V / 100 mA microcontroller programming (through its USB connection to the computer) without having to feed on the other hand. However, it mounted but not sold in kit form, so a minimum welding of components is required (though the instructions provided are clear and detailed). The installation of a driver is also required if you use the Windows operating system on our computer.

"USBasp" (http://www.fischl.de/usbasp): ISP programmer scheme whose hardware and firmware is free. It requires installation of a specific driver in Windows operating systems.

"Arduino as ISP": if we have two Arduino boards and want to program the microcontroller of one of them, we can connect them to (communicating via SPI) the first microcontroller programmer function as the second. If you want more details on page http://arduino.cc/en/Tutorial/ArduinoISP is explained step by step the whole procedure.

Other ISP programmers noteworthy are the "USB AVR Programmer" Pololu (product # 1300) and the "USB AVR Programmer" of Seeedstudio (TOL132C1B product code).

There are two standards for ICSP connectors, one 6-pin and a 10, but the latter is obsolete. The Arduino UNO incorporates a ICSP connector
6 pin whose schematic diagram is shown in the following figure.

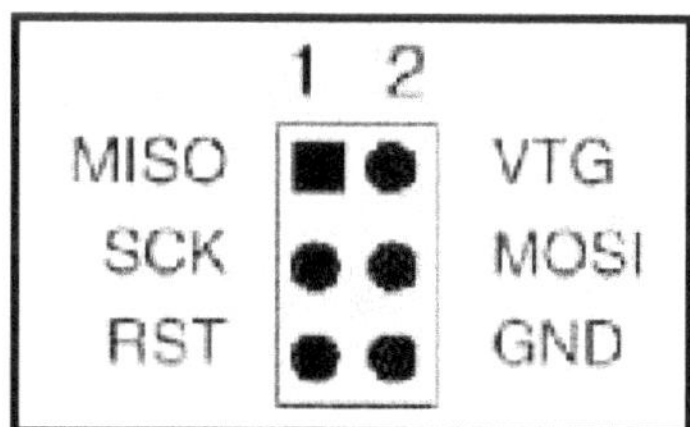

Each pin ICSP connector is spliced internally to a specific pin of microcontroller ATtmega328P. From the above figure we can see that the ICSP connector actually uses the standard SPI protocol to communicate with the microcontroller programming. Specifically, we have

besides the power supply pin (VCC) and ground (GND) - The clock pin (SCK,

- "Clock" - that sets the rate at which data is transferred), the output pin series

data (MISO), pin serial data input (MOSI) and called RESET pin (SS pin connected to the microcontroller). The function of the latter pin is enable or disable communication with the microcontroller: while a high voltage nothing happens is received, but when a low voltage, ATmega328P stop program execution have recorded at that time received and disposed to receive a rescheduling.

If we had acquired a microcontroller as spare part and we had no Arduino with free DIP socket for placement in order to program it using ICSP we need a ICSP connector where plugging our programmer ISP and connect each pin of the connector with the corresponding pin on the microcontroller. It is relatively simple to manufacture the relevant circuit by a breadboard. Even if you have sufficient knowledge, it is not difficult manufactured a small side plate perfboard type with connections established between connector and microcontroller. Examples of such plates (along with a detailed explanation of manufacture) can be found on the site http://www.evilmadscientist.com. You can also purchase an exclusive model Sparkfun, with product number 8508.

Anyway, not enough to properly connect your ISP programmer to program our microcontroller using ICSP. You must perform a series of steps involving the use of certain software running on our computer. However, the specific procedure can not explain it yet because we still lack the necessary knowledge, but will be detailed in the next chapter.

The clock

To mark the pace of execution of the instructions in the microcontroller, the pace of reading and writing data on their (s) report (s), the rate of data acquisition input pins, the rate shipping data to the output pins and generally to control the frequency of the microcontroller, the Arduino board has a small "metronome" or clock, which operates at a frequency of 16 million Hertz (16MHz). This means that (About and oversimplify) the microprocessor is capable of performing
16 million instructions per second. Could the Arduino incorporate a clock with a frequency of greater work, and decrease the time in which the instructions are executed, but this would also mean an increase in power consumption and heat generation.

Electronically speaking, there are several types of "clocks". Among them are crystal oscillators and ceramic resonators. The former are circuits using a piezoelectric material (normally quartz crystal, hence the name) for generating a vibratory very precise high frequency wave. The latter consist of a piezoelectric ceramic material that generates the oscillating signal of the desired frequency is applied when a given voltage.

The watch bearing the Arduino is a ceramic resonator. Such clocks are slightly less accurate than crystal oscillators, but are cheaper. Specifically the frequency accuracy provided by a quartz crystal with temperature compensation (called TCXO) is 0.001% (i.e., a nominal value of 16 MHz would have values as much of 16 KHz above or below this), but the precision provided by a typical ceramic resonator made of PZT (lead zirconate titanate) is 0.5%.

If we are not satisfied with the accuracy of the clock that incorporates the Arduino, we can always choose to use an external component to act as different from what is in the plate itself and synchronize the microcontroller with the clock, but in fact, for the vast majority of Arduino projects we do, we will not need to use this for anything.

Technically, one could use as ATmega328P clock crystal oscillator present on the plate controlling the ATmega16U2, but in practice the resulting circuit would generate too electromagnetic noise.

On the other hand, it is worth commenting that the ATmega328P microcontroller actually includes its own internal clock (RC type) within its encapsulated, so in theory there would be no further need to use the Arduino clock. However, the internal clock is only able to dial a "rhythm" of 8 MHz and also has a very poor accuracy (10%) so it is almost never used.

The button "reset"

The Arduino UNO has a reset button ("reset") that allows, once pressed, send a signal to the pin LOW "RESET" of the plate to stop and restart the microcontroller. As at the time of starting the execution microcontroller bootloader always activated, the reset button is typically used to allow loading of a new program in the Flash memory of the microcontroller-eliminating which was recorded earlier-and subsequent commissioning up. However, the plate UNO is not necessary practically never press "real and physically," the button before each load, since the plate ONE is designed in such a way that allows the activation of the bootloader directly from the development environment installed on your computer (simply by clicking on an icon of the environment).

This ability plate "reset" without physically pushing any button has its consequences: when the plate ONE is connected to a computer running a Linux or Mac OS X, every time a connection is made via USB plate This restarts, so for about half a second, the bootloader starts. Therefore, during the second half, the bootloader will intercept the first few bytes of data sent to the plate just after setting the USB connection. The problem with this is that although the bootloader will ignore these bytes because it ignores these no longer arrive at the desired destination (ie, the recorded program in the microcontroller itself), because this will start with the delay half a second. Therefore, we must bear in mind that if our recorded in the microcontroller program is designed to receive data from the very start of our computer via USB, we have to ensure that the software responsible for sending a second wait after open the USB connection to proceed to send.

If desired, you can disable the "auto-reset", so that it is always necessary to use the built-in board to run the bootloader button and therefore load a new program into the microcontroller. There are several ways to achieve this, but the simplest is to connect a 10 microfarad capacitor between pins "GND" and "RESET" of the plate. Another more definitive to remove the "auto-reset" way would be to cut the trace labeled "RESET-IN" on the back of the plate; to regain the "auto-reset", each end of the trace should be welded back together. In any case, once disabled the "auto-reset", the charging process of our programs is as follows:

1. Hold down the reset button on the board.
2. Click on the "Upload" button Arduino development environment.
3. As soon as the LED lights once labeled as RX in our
Arduino, quickly release the button.
4. The burden should begin by flashing the LEDs RX and TX.

Get the schematic design and reference

Curious (with an advanced level of electronic) who want to learn to detail how it is constructed Arduino UNO and how interconnected the different components that form, schematic design can be downloaded in pdf format from the following address: http: //arduino.cc/en/uploads/Main/Arduino_Uno_Rev3-schematic.pdf

Also available reference design needed to build ourselves a printed circuit which is exactly equal to that of the official Arduino (or if you modify these files modified) and then supplement it with the necessary components (microcontrollers, female-PIN, etc.) separately. This design can be downloaded from the following address: http://arduino.cc/en/uploads/Main/arduino_Uno_Rev3-02-TH.zip. If you unzip the zip file, you will see that contains two files: the file .SCH is a schematic diagram visual PCB containing a series of graphic symbols representing mainly logic gates and connecting lines whose usefulness is easily generated from the file it .BRD, which is what actually contains all data necessary for the manufacture of the PCB.

The contents of these files can be viewed and edited using a software (not free) PCB design called EAGLE-version 6.0 or later date. You can download it from their official website http://www.cadsoftusa.com A free 30-day trial.

Clarify that a microcontroller that is not carrying the Arduino (specifically, it is the microcontroller ATmega8) appears on these design files, but there is no problem with this because the pin configurations microcontroller ATmega8 (and ATmega168 too) They are identical to those of the ATMEGA328P.

Finally say that the Arduino is certified by both the US Federal Communications Commission (FCC) and the European Union. That means Arduino meets in both geographical areas with the relevant regulations in the field of electromagnetic emissions of electronics

and telecommunications. We can verify that indeed Arduino respects the rules of those organizations seeing their logos on the back of the plate. The importance of having these printed logos is to facilitate the acceptance of Arduino boards in their access to different markets (US and Europe as appropriate) as if they were not certified, entry may be rejected by not complying with legal regulations.

WHAT ARE OTHER OFFICIAL Arduino boards?

Here is a brief summary of the possibilities offered by various other official Arduino UNO model plates. All these variants are specialized in working within specific circumstances where plaque ONE standard does not offer solutions to the needs that we may arise. Anyway, if you want to get more detailed and comprehensive information than is provided in the following paragraphs, it is best to consult the following link: http://arduino.cc/en/Main/Products, which specifies all technical details of each of these plates.

Arduino Mega 2560

Microcontroller board based on the ATmega2560. As main features we say that has 54-pin digital input / output (of which 14 can be used as PWM analog outputs), 16 analog and 4 receiver / transmitter UART TTL-series entries. Flash memory consists of a 256 Kilobytes (of which 8 are reserved for the bootloader) SRAM memory
8K and 4K EEPROM. Its working voltage is equal to the model
ONE: 5V

The entry for this plate section within the official website of Arduino we we can download the files from the schematic design of the plate in PDF files of the PCB layout in the format of the EAGLE program and an illustrative mapping pins microcontroller relative to the pins of the plate. We can also download the official documentation ATmega2560 microcontroller.

Arduino Mega ADK

Mega very similar to plate 2560. The main difference is that the Mega ADK is able to function as a device type "USB host" (and not Mega 2560). Let us explain this.

In a USB communication between different devices there is always one that acts as a "master" (the so-called "host") and the other -u others that act as "slaves" (called "peripherals"). The "host" is the only one who can initiate and control the transfer of data between other connected devices, while "peripheral" can only respond to requests made by the "host" and little else. The "host" devices have a USB type A connector, and the "peripheral" devices have a USB type B connector, Mini-B or Micro-B. A typical "host" is a computer, which can be connected to several "peripheral" as memory sticks, cameras or video, next-generation mobile phones, etc.

Arduino ADK can function as USB peripherals like the rest of Arduino boards (thus incorporates the USB connector type B and others) but also as a USB host (and so, as a novelty, also incorporates a USB connector type TO). But it is not enough to have the proper connector to the plate can be a USB host: actually this mode of operation is possible because the plate incorporates a specific chip (microcontroller, in fact) that implements the necessary logic: the Manufacturer Maxim MAX3421E, which communicates with the ATmega2560 via SPI.

Thus, the ADK board can connect any device with a peripheral USB port (mobile phones, cameras or video, keyboards, mice, joysticks and controls from different consoles, etc.) to control and interact directly with it.

Specifically, the Arduino ADK is specially designed to interact with mobile phones running the Android system (http://www.android.com), developed by Google. The idea is that you can write programs for Android that relate to the Arduino code executed at that time on the plate, so that a communication between the mobile and the plate to allow, for example, make a remote control established from Android device sensors and / or actuators connected to the Arduino hardware.

The combination of Android Arduino is a tremendously interesting but very large and relatively complex issue (need to know how it works internally platform Google, in particular, the development kit "Accesory Dev Kit"), so in this book will not be addressed. If you want to know more, please consult the final bibliography of the book provided or start here: http://labs.arduino.cc/ADK/Index. You can also consult the official documentation of Android, especially the sections corresponding to the communication path
USB: http://developer.android.com/guide/topics/usb/adk.html and http://developer.android.com/guide/topics/usb/index.html

Arduino Ethernet

As the model ONE, Ethernet board is based on the microcontroller ATmega328P (and therefore has the same amount of Flash, SRAM and EPPROM memory), and also has the same number of pins of digital input / output and analog inputs. All other features are also very similar to the model ONE. The biggest difference with the plate One is that the plate incorporates an Ethernet RJ-45 socket type to connect using the appropriate cable (twisted-pair Category 5 or 6) to an Ethernet network.

Ethernet networks based on the IEEE802.3 specification (http://www.ieee802.org/3) are the most widespread technology far in building local area networks (called "LAN" -Local Area Network -). LANs allow data traffic between devices connected to it (computers, printers, "routers", etc.) within an area of relatively short distances, such as a building or a set of adjacent buildings. Although technically not be why, in practice Ethernet networks use the same set of communication protocols that Internet (the call stack "TCP / IP").

The Arduino Ethernet board therefore allows transfer data between itself (which can be obtained from a sensor, for example) and any other device connected to the same LAN (usually a computer that collects and stores), or vice versa : transfer data between a device connected to the LAN (typically a computer running some software control) and herself (which may be connected to a computer remotely controlled by the actuator). You can also achieve, by establishing appropriate routing packets, communicate our Ethernet card with any device connected to any network in the world outside our private LAN (including "Internet"), bringing its usability soar.

Obviously, only to have an RJ-45 socket the Arduino Ethernet board can not "magically" to communicate within a network. Allowing this to happen is the inclusion within the plate chip Ethernet controller, in fact, all hardware implements the TCP / IP protocols (specifically TCP, UDP, ICMP, IPv4, ARP, IGMPv2, and PPPoE
Ethernet course). This is the W5100 Wiznet manufacturer. This chip is the real responsible for the Arduino Ethernet board can handle such networks.

This chip "hardware implements the TCP / IP stack" means in a nutshell that allows the Arduino software developer very easily use the network (by programming library "Ethernet", built by default within the Arduino language) to to transmit or receive data without having to worry about technical details (such as error control in data transmission, synchronization signals and datagrams, etc.) of which already this chip is directly responsible. On the detailed library "Ethernet" programming (including basic networking concepts such as configuring IP addresses, MAC addresses, etc.) and their possible applications discussed in Chapter 8 use.

Among other features, it notes that the W5100 can operate at transmission speeds of 10 and 100 megabits per second, which allows up to four independent simultaneous connections and has an internal memory of 16 kilobytes to temporarily store data sent or received from the network . The W5100 communicates via SPI protocol with ATmega328P, so keep in mind that the digital pin numbers 10, 11, 12 and 13 of the Arduino Ethernet board reserved (ie, can not be used for anything else). For this reason, the pin that is associated with the LED that is built into the motherboard is not the number 13 as with the plate UNO, it is number 9, plus the number of pins available is reduced compared to the model ONE 9 with only four pins available as PWM outputs.

The Arduino Ethernet board also has a socket for inserting a microSD card, which can be used (by programming library "SD", built by default within the Arduino language) to store different types of files and offer them through network (remember, can be our private LAN or, if we want the whole world). Keep in mind that if the microSD card is present, the pin 4 is reserved for the control of this.

USB-Serial adapters

What you might call the attention of the Arduino Ethernet board is that it has no USB socket (the "price to pay" to have instead of an RJ-45 socket). This means that from our computer can not communicate via USB (the "standard" form) with the bootloader

our microcontroller so that we can record in its Flash memory the program we want to run. To resolve this obstacle, we could use an ISP programmer, but the simplest is to acquire and use the official USB-Series (http://arduino.cc/en/Main/USBSerial) adapter. This adapter is merely an insert containing a mini-USB socket type B and microcontroller ATmega8U2, very similar to the already known ATmega16U2 chip (but with half of Flash memory, hence the name) that is programmed to perform exactly the same function: converting the USB connection in a simple series of 5 V signal understandable by the microcontroller ATmega328P Ethernet board.

Therefore, to connect via USB to Ethernet board we plug first this adapter to the six pin protruding from the Ethernet card (each of which is in turn connected directly to a particular pin of ATmega328P: RX, TX, reboot, power, ground ...) and then connect the USB cable to the socket provided by the adapter. As we do with the plate ONE, the USB connector can use the newly added thereafter both electrically power the board to program it (even with self-reset built too).

However, if you do not want to use the official adapter, many other adapters that perform the same function, but generally feature a different USB converter chip-Series: the manufacturer FTDI FT232RL. An example of these adapter plates are called "FTDI Basic Breakout" distributed Sparkfun (which according to the model can operate at 5 V or 3.3 V) or official adapter cables also "USBTTLSerial" FTDI (namely TTL models -232R-5V and TTL-232R-3V3, which, as its name suggests, operate at 5 V or 3.3 V, respectively). Similar plates are the "FTDIFriend" Adafruit, the "FTDI Adapter" of Akafugu or "seal" of IteadStudio, among many others.

PoE ("Power over Ethernet")

Something you should know regarding the Arduino Ethernet board is that it is possible to electrically feed itself through the Ethernet cable, unused or USB cable or external power supply (AC / DC, battery, etc.). That is, you can benefit from free data offered by the typical cable for an Ethernet network there also needed for the proper functioning of the plate voltage, without therefore using other twisted pair cable. This is very convenient in facilities whose size or location made complicated the addition of several cables.

For this, however, they must meet different conditions. The first is through the Ethernet cable properly transformed trip signal. This can be achieved in two ways:

Using an injector "midspan". This is a device that plugs on one hand both the external power supply (via AC / DC adapter, usually) as a standard network switch (using an Ethernet cable), and on the other with the device to be fed (by just another Ethernet) cable. The data pass through the injector to / from the switch or from / to the end device by Ethernet cables both sections unchanged, but also the power received by the injector will be transmitted by the Ethernet cable connected to the end device. Basically the scheme described would be the following figure:

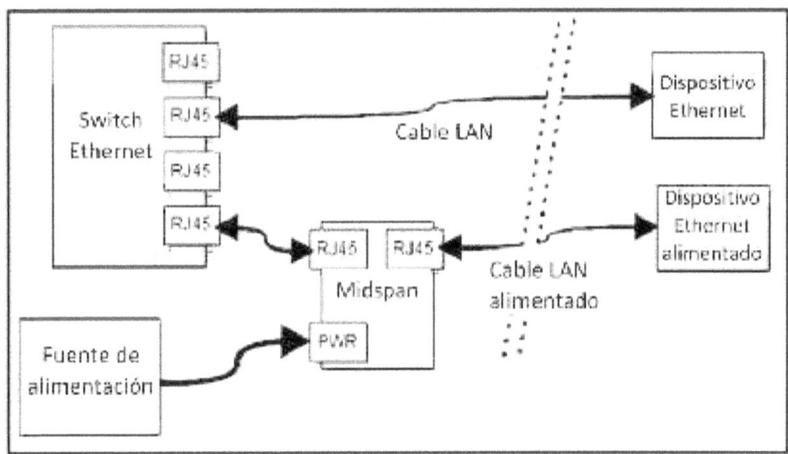

Using a PoE switch. This is a switch (a network switch) special addition to functioning as a standard Ethernet switch (intercommunicating together the devices connected to it) has the ability to provide power through their sockets RJ
45. Therefore, we want that both devices are connected to each other via Ethernet as they are electrically powered via PoE must plug into this type of switch, using any standard Ethernet cable.

Unfortunately, the first PoE (both as switches PoE midspan injectors) systems were created by several companies without following any particular standard, so there are different incompatible systems PoE (since They use different voltages, different polarities, various provisions of pins, etc.). Fortunately, a standard specification was finally defined and PoE current systems have converged to reach a point where they are almost compatible. This specification is called "802.3af" (2003) or the more modern, "802.3at" (2009).

However, these standards are too difficult to implement because stipulate a voltage value to use up to 48 V (voltage more than the components used in our projects can handle, even for voltage regulators) and a relatively signaling scheme complex feed device has implemented to inform the injector on the amount of tension required. Therefore, often it tends to use a simplified system that uses the same connections as the 802.3 standard but works at a lower voltage, without signaling scheme.

In other words, to use with Arduino PoE is not necessary to use a switch or injector "midspan" that supports the full 802.3 standard, which also happen to be quite expensive. We can simply use an injector "midspan" as basic as the so-called "Passive PoE Cable Set" Sparkfun (shown at left), which has just one side of a connector "jack" female 2.1 mm for connecting with the power supply and an RJ-45 connector
to plug in a switch, and the other has an RJ-45 socket where the Ethernet cable to carry the signal power PoE device is placed. SparkFun sells along with a PoE splitter (we shall see what this is) with product code
10759.

Other more sophisticated but also very cheap injector "midspan" and easy to use is the so-called "4-Channel PoE Midspan Injector" manufactured by Freetronics, which provides a voltage up to 24 V (to spare for what an Arduino is able to handle). Specifically, the injector has a power connector "jack" 2.1mm prepared to be connected to a power supply that provides a voltage of between 7 V and 24 V (AC or DC), 4 sockets RJ-
45 intended to be connected to a standard network switch and 4 RJ-45 sockets
intended to be connected to different feed devices. This power is held by an output voltage of the injector which will be identical to the input value but always DC. Therefore, according to the value of the voltage applied to

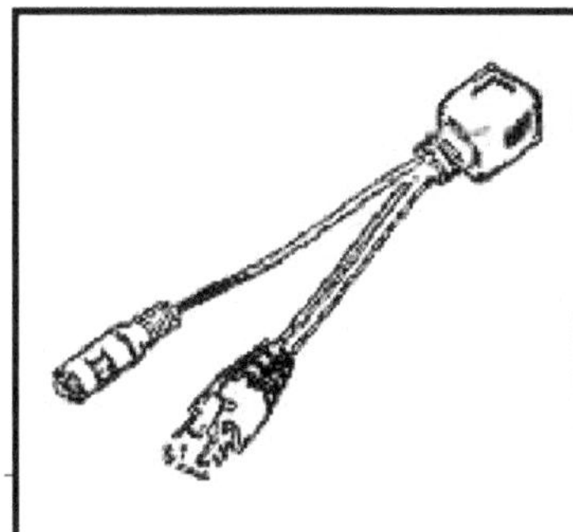

injector, it may be necessary coupling between this device and a voltage regulator to reduce it to more acceptable values for our Arduino boards.

The second condition to be met in order to use the PoE is fed (and connected at a time) device for the Ethernet cable has the capacity to correctly process the PoE signal that you get in the cable. If the device already is enabled to handle signals PoE, no more to do: simply connects the Ethernet and ready cable. If that device, however, is not able to interpret the signals PoE, it just means the data communication but can not receive power. To achieve this, it can be done in several ways:

Use a "splitter PoE". This is simply a cable that on one side has an RJ-45 socket which is inserting the Ethernet cable that transmits the PoE signal and the other is divided into two connectors to plug the device: an RJ-45 that transmits data and a connector "jack" of 2.1 mm, which transmits DC. The following figure shows an example of "PoE splitter."

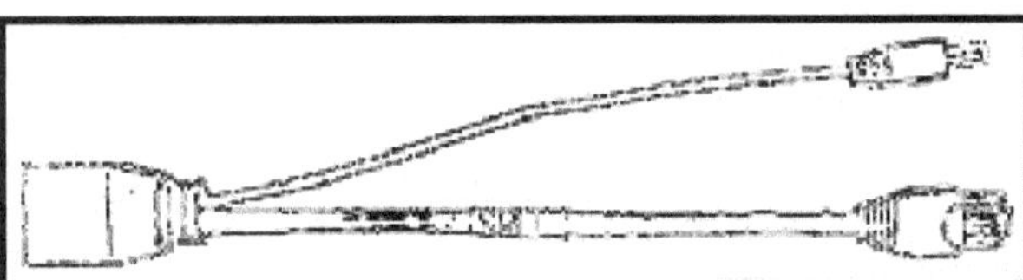

Attach an extra small plate (also called "module") to the device in question, that it provide this capacity PoE conveniently process the signal received via the Ethernet cable. In the case of the Arduino Ethernet board, you can purchase the PoE Ag9120-S module, specifically adapted to it, which offers 9 V output to the plate with a range of input voltage from the LAN cable 36 V and 48
V. However, this is not officially module manufactured by the Arduino
Team because it's proprietary hardware. It can be purchased either on the websites of the various distributors listed in Appendix A, or on the manufacturer's own website (http://www.silvertel.com/poe_products.htm).

Arduino Fio

This plate contains a ATmega328P operating at 3.3 V and 8 MHz. It has 14 holes that can be used (by direct welding or by placing plastic female pin-) as input pins / digital output (6

which can be used as PWM output); also it has 8 holes prepared for use as analog inputs and a reset button, all within a very small size.

A new feature of this board over previous is that you can feed electrically by a LiPo battery through the motherboard has a socket type JST 2 pin to connect directly there. Arduino Fio can also be powered via USB, because it has a USB mini-B connector for it. In fact, through the power received via USB you can even recharge the LiPo battery is connected at the time, since the plate incorporates the charger manufacturer Maxim MAX1555 chip. However, the USB connection is not intended to program the microcontroller, so this requires attaching a USB-Serial adapter (as already mentioned when we saw the Arduino Ethernet board) to the holes of the plate marked GND, AREF , 3V3, RXI, TXO and DTR through the string of adequate pin.

Anyway, the most interesting novelty of this board is the ability to put a XBee module into the socket incorporates specific to it. "XBee" is the trade name given by the manufacturer Digi International to a family of transmission / receiving radio frequency signals with low power consumption and an encapsulation and size compatible with each other. So Arduino Fio is intended for wireless applications that are autonomous in their functioning and therefore not requiring a high level of maintenance. A very common case is connected to this plate a sensor of any kind with the XBee module to create several plates Fio a wireless network that allows these sensors interrelate with each other and with a central computer data collector. The XBee modules, however, not be discussed in this book.

Arduino Pro

This board comes in two "versions": both contain a ATMEGA328P SMD microcontroller, but operates with 3.3 V and 8 MHz and the other works with 5 V and 16 MHz has 14 holes designed to function as input pins /. digital output (6 of which can be used as PWM output), 6 analog inputs holes, holes for mounting a power connector 2.1 mm, JST socket for an external LiPo battery, a power switch, a button Restart a ICSP connector and the pins needed to connect an adapter or USB-Serial Cable so you can schedule it (and feed) directly via USB.

This board is designed to be installed semi-permanently in objects or exhibitions. Why it not come with the pins mounted but must be placed in the holes the plastic pin-female "hand" (or solder wires directly). Thus, using different types of configurations it is allowed as required.

Arduino Lilypad

LilyPad plate is designed to be stitched fabric. It also allows connect (through wires) power supplies, sensors and actuators so that they can "carry", making possible the creation of clothing and "smart" clothing. In addition, you can wash. This board incorporates the microcontroller ATmega328V (a version of low consumption of ATMEGA328P), which is programmed by coupling to the plate adapter or USB-Serial Cable.

In http://www.sparkfun.com/categories/135 link can be found several interesting supplements adapted in size and flexibility to LilyPad such as temperature sensors, and accelerometers light, LEDs of different colors, vibration motors , buzzers, wireless transceivers, LiPo battery holder button or AAA, breadboards, switches, buttons, specific variants of the plate, etc. Even sets of these supplements are distributed kits called "ProtoSnaps" for comfort.

On the other hand, at the official store of Arduino (see Appendix A) can acquire the "Wearable Kit" consists of a set of components (resistors, potentiometers, breadboards, thread ...), a pressure sensor and actuator several (buttons, LEDs ...), all textiles. This kit is designed by the specialist company Plug & Wear.

To learn more about potential uses and tricks of this board, you can view the tutorials No. 281, No. 308, No. 312, No. 313 and No. 333 of Sparkfun, and also comprehensive web practice and co-designer, Leah Buechley http://web.media.mit.edu/~leah/LilyPad. If you want to know projects and
made with these components, a good place for inspiration is http://www.kobakant.at/DIY

Arduino Nano

The most outstanding feature of this board is that despite its size
(0.73 inches wide by 1.70 long), still offers the same number
of digital and analog outputs and the Arduino UNO and the same functionality as the inputs. The most obvious consequence of small size is that it lacks the power connector of 2.1 mm (but can still be fed by an external source via pin "Vin" or "5 V") and features a USB mini-B connector instead of USB connector type B.

Another difference is that, although Arduino Nano is still based on the ATmega328P microcontroller (in SMD format), the USB-serial converter that incorporates the FTDI FT232RL chip is not the ATmega16U2.

This board is especially designed to connect to a breadboard using the pins protruding from his back, and may form part of a complex circuit of a relatively fixed manner.

Arduino Mini

This board is very similar to the Arduino Nano is also based on SMD ATmega328P microcontroller running at 16MHz, it has 14 pins for digital input / output (6 of which can function as PWM outputs) and 8 analog inputs. And like the Arduino Nano, the Arduino Mini is specially designed to connect to a breadboard using the pins protruding from his back, and may form part of a complex circuit of a relatively fixed manner.

The most important difference Arduino Nano is that the Arduino Mini (to save even more space and thus achieve a truly minimum size of 0.7 inches wide by 1.3 long) does not incorporate any USB- converter chip series. As a result, for programming is required to use an external USB-Serial adapter. Specifically, the use of one specific and official recommended: the so-called "Mini USB" based on the FTDI FT232RL (http://arduino.cc/en/Main/MiniUSB) chip, which should be placed separately in the breadboard and then establish relevant links between this and the Arduino Mini by cables. For full details of this specific configuration, and overall experience different possible uses of this board, I recommend consulting the official guide: http://arduino.cc/en/Guide/ArduinoMini.

Arduino Pro Mini

This board has the same size as an Arduino Mini, and a compatible pinouts. It comes in two "versions": both contain a
ATmega168 but operates with 3.3 V and 8 MHz and the other works with 5 V and 16 MHz. It also features a reset button and the pins needed to connect an adapter or USB-Serial cable so you can schedule it (and also feed) directly via USB. It can also be fed by an external source electrically connected to the pin "Vcc".

This board is designed to be installed semi-permanently in objects or exhibitions. Why it not come with the pins mounted but must be placed in the holes the plastic pin-female "hand" (or soldering wires directly). Thus, using different types of configurations it is allowed as required.

Arduino Leonardo

The novelty of this board is the microcontroller that incorporates is the ATmega32U4 (SMD format), which has all the features that the ATMEGA328P but also incorporates more than 0.5 kilobytes SRAM and above, supports USB communications directly (and therefore does not need any additional chip as ATmega16U2 or FTDI).

Other differences with the board One is that Leonardo board incorporates a digital pin-female more than the one to be used as PWM output (# 13) and added extra 6 analog inputs, which are physically located in the digital-PIN female marked with a dot on the outside of the plate.

Another difference is that the SDA and SCL pins for I2C / TWI change location about UNO and now happen to be on pins-female digital No. 2 and No. 3. Furthermore, in the Leonardo disappear GPIO pin plate SPI, so the only way to communicate with the outside this plate using this protocol is directly using ICSP pin.

The fact that only Leonardo plate both include a microcontroller to implement programs to communicate directly via USB to the computer allows this board can easily simulate (if properly program) be a keyboard or a USB mouse connected to said computer. Technically, when the Leonardo plate is connected with a USB cable to the computer, this will detect two "ports" of different connection: a ready to use the board as a peripheral Leonardo USB plus standard USB port (typically would be a keyboard or a mouse, as we said) and a different port, similar to that generated when the Arduino UNO is connected, usable in "traditional" way to
programming and communication with the board through the programming environment Arduino.

The "auto-reset" of Leonardo plate

Also because the board uses a unique Leonardo microcontroller for both implementation of programs to the USB communication with the computer, to restart the microcontroller (physically pushing the button "reset" of the plate or via the corresponding button development environment) USB connection similar to the UNO (the other not) is interrupted and then restored. Not so with the plate ONE USB connection because there always remains a separate chip (the ATmega16U2) that is never restarted. One consequence of this is that any program you are communicating at the time through a serial connection via USB to lose its connection plate Leonardo.

Another consequence is that the burden of the programs carried out by the bootloader will be delayed a few seconds (usually eight) and that after activating the restart, the Arduino development environment should wait until it detects again an active USB connection and then make the program load through it. If the physical button "reset" is used, this delay means that we keep pressing this button during those seconds until you see the message "Uploading ..." on the bottom bar of the development environment: you can not let go before.

Technically, the "auto-reset" of Leonardo plate is activated when the computer sends a signal series of 1200 bits / s and closed. This means that at the start of the plate will be no delay in the execution of the recorded program, because if that signal is not received, the bootloader will not run.

Arduino Micro

This board offers the same features as the Arduino Leonardo (has for example the same microcontroller ATmega32U4 to 16MHz, the same 32KB of Flash memory and 2.5 KB of SRAM, the same bootloader, the same working voltage -5 V ...) but with really minimum size: 48 x 18 mm, ideal for placement on a breadboard without taking up much space. Like the Leonardo model can be programmed via a USB connection (has a mini-B to this socket) and can also function as simulated keyboard or mouse. The characteristics of the "auto-reset" of Leonardo board also apply to the Micro. It can be powered via USB or cable by an external power supply connected to pins "Vin" and "GND".

Arduino Due

This plate belongs to a completely different from that of other family Arduino boards. Includes SAM3X8E microcontroller, which, although manufactured by Atmel is a very different AVR internal architecture (ie is of type ARM Cortex-M3) and also its records are four times larger than usual in the other plates (specifically, they are 32 bits). Its clock speed is also well above the rest of Arduino boards (specifically, it is 84 MHz). In addition, the microcontroller SAM3X8E has many more memory (ie, 96 KB of SRAM and 512 KB of flash memory) and a dedicated circuit (called "DMA" controller) which allows the CPU to access the memory of a long way faster.

All this implies that with Arduino Due can do more, and faster, so lets run applications that perform a large data processing. The price is also higher than the rest of Arduino boards, of course.

Other technical data of the Arduino Due are available 54-pin digital input / output (12 of which can be used as PWM outputs), 12 analog inputs, 4 chips TTL-UART (ie, four channels independent hardware series) , 2 digital-analog converters (novelty!), 2 independent I2C, SPI port 1 (which is only implemented on pins "ICSP") ports, 1 USB mini-B connector type, 1 USB type mini-connector A, a socket type 2.1mm "jack", a reset button and a delete button. It offers as usual pins "Vin", "GND", "5 V" and "3.3 V".

A very important aspect of this board you need to know is that their working voltage is 3.3 V. This means that the maximum voltage pins I / O can withstand that is. If provided with a higher voltage (such as 5 V to which we are accustomed), the plate could be damaged. However, the external power may be the same as those used with the Arduino UNO as its input voltage ranges are identical (6-20 theoretical V 7-12 V recommended); This is because the tension is properly reduced thanks to an internal regulator. On the other hand, the intensity offered by the output pins is between 6 mA and 15 mA, and that offered by the pins "3.3" and "5 V" is
800 mA.

The Arduino Due maintains the shape and arrangement of the Arduino Mega, is compatible with all shields respect the same pinout and, importantly, work to 3.3 V. On the other hand, all pin input / output have a resistance "pull-up" internal default disconnected 100k.

The Arduino Due offers two USB connectors to separate two different functionalities. The closest to the power jack (mini-B) connector is designed to plug the board into the computer and transferred from the development environment our program to be executed by the microcontroller, and from there keep serial communication between computer and board . In fact, this connector is controlled by the same chip ATmega16U2 the Arduino UNO board so its behavior is identical. Nearest the reset button (mini-A), however, the connector is directly controlled by the SAM3X8E chip and is intended for use as a peripheral USB plate more (as a keyboard or a mouse, as also it occurs in the Leonardo plate). In addition, a novelty that provides the latter connector is that it also allows the board to act as "USB host" (as does the Mega plate). Thus, not only could we use the Due plate as a keyboard or a mouse, but it could connect a keyboard or a real mouse, among many other devices such as mobile phones of last generation, for example.

How to program the Arduino Due is similar to previous plates, both in the use of the development environment and the programming language itself: all changes are "under the surface". However, the current version of the development environment to date edition of the book (1.0.2) does not yet allow the use of this plate, so you need the version 1.5, which at the time of issue This book is in beta. The download can be done from here http://arduino.cc/en/Main/SoftwareDue.

The only detail to bear in mind is that the Flash memory of the microcontroller must be removed "manually" every time you want to load in our program. This is because the bootloader of this board is housed in a separate memory Flash ROM type memory, and runs only when it detects that the Flash memory is empty. Hence the existence of the delete button (marked "Erase") on the plate. Fortunately, if we connect to our computer via USB mini-B socket (the closest to the power jack), this deletion process is performed automatically. In this regard, the USB communication between board and computer is performed in the same way using either Due plate (through the mini-B connector) as the plate ONE.

WHAT "SHIELDS" ARDUINO OFFICIAL THERE?

Besides Arduino plates themselves, there are also called "shields". A "shield" (in English means "shield") is simply a printed circuit board that is placed on top of an Arduino board and connected to it by engagement of the pins without any cable. Its function is to act as extra plates, extending the capabilities and supplementing the core functionality Arduino in a more compact and stable.

Depending on the model, even several shields can be stacked one atop another. This will depend on whether the lower shield provides pin-female to attach them to pin back protruding upper shield.

Normally, shields share the GND, 5 V (or 3V3) and AREF RESET lines with the Arduino, and also tend to monopolize the use of some pins input / output for their own communication with her, so these are " unused "for any other use. For example, if several shields engage each other and all communicate via SPI with the Arduino, all of them can be used without problems corresponding to the lines MISO, MOSI and SCLK pin common, but each of them must use a pin different as CS line.

On the other hand, we must also consider the power requirements needed by shields. We know that an Arduino board receives about 500 mA (either via USB or by connecting external jack), so the current is for the operation of a possible shields is small. Examples shields consuming (up to 300 mA) are those LCD screens that provide connectivity or Wi-Fi. We must also take into account whether a certain shield requires a voltage of 3.3 V.

There are literally hundreds of shields built by the community with the support plate one that will bring an added versatility, but "shields" are only the following officers:

Arduino Ethernet Shield

This shield is intended for those who want to add to the Arduino UNO the ability to connect to a TCP / IP wired network. It provides the same functionality as the Arduino Ethernet board but complementary shaped shield coupled to the Arduino UNO. In fact, this shield has the same controller chip that W5100 Arduino Ethernet board, and configured with the same programming library. The "Ethernet", which now comes standard Arduino language bookstore

Once connected to the shield plate through the string ONE pin which fits perfectly up and down, to our circuits thereafter use the inputs and outputs offered by the pin-female Ethernet shield. These inputs and outputs have exactly the same layout and functionality as UNO plate. Even if necessary, it could seamlessly connect one second shield on top of Ethernet shield to keep adding functionality.

The process of loading our programs in the UNO microcontroller plate coupled to Ethernet shield unchanged compared to normally done with a plate ONE freelance must first connect the board UNO to our computer via the USB cable, and once the program is loaded, as always, we will continue to feed the plate via USB or disconnect it from the computer and plug it into the external power supply. Thereafter, the RJ-45 connector shield we can connect a network cable (technically, a twisted-pair Category 5 or 6) "standard" type if we communicate the shield to a switch or router or type "crossed" if we want to communicate the shield directly to a computer.

This shield requires 5 V to operate. This voltage is provided by the UNO plate by fitting the corresponding pin power between plate and shield. Communication between the chip and the plate W5100 ONE is set by pins
10,11,12 and 13 (via SPI) so that these pins can not be used for other
purpose. This means that when using this shield really have 4 digital inputs / outputs less.

This shield also includes (as does the Arduino Ethernet board) a socket to place a microSD card, which can be used by programming library "SD" which comes standard in the Arduino language. As is the case with W5100 chip to communicate with the microSD card Arduino UNO uses the SPI protocol, but this time it uses as SS pin number 4 instead of 10 (reserved to W5100). That is, used pins 4, 11, 12 and
13.
Sharing the SPI channel between the W5100 chip and microSD card means that you can not use these two devices simultaneously. If we use in our programs both elements, we consider this to program your code correctly. If, however, we do not use one of the two devices must be taken to disable
explicitly in your code; to clear the microSD card has been configured as an output pin 4 and writing a HIGH value and to deactivate -ALTO- W5100 chip has to do the same but with the pin 10.

Like the Arduino Ethernet board, the shield also has the possibility of acoplársele a PoE module.

Finally, indicate that this shield also has its own button "reset" which resets both the W5100 and the Arduino chip itself, and a number of interesting information to know LEDs "PWR" (indicates that the board and the shield are powered), "LINK" (indicating the presence of a network connection, and flashes when the shield transmits or receives data), "FULLD" (indicating that the network connection is "full duplex"), "100M "(indicating the presence of a network connection of 100 megabits / s -instead of one of 10 megabits / s-)," RX "(flashes when the shield receives data)," TX "(flashes when the shield sends data) and "COLL" (flashes when packet collisions are detected on the network).

Arduino Wireless SD Shield

This shield is designed to allow an Arduino UNO able to communicate wirelessly using a XBee module (purchased separately) or similar plate. This establishes a link with other XBee device at a distance of up to 100 meters indoors and up to 300 meters outdoors in line of sight.

As with the rest of official shields once connected the shield plate on the UNO through the string of pins that fit perfectly above and below, to our circuits thereafter use the inputs and outputs offered by the pin-female This shield. These inputs and outputs have exactly the same layout and functionality as UNO plate and even if necessary, could be connected seamlessly one second shield on top of the shield.

This shield has a switch labeled "Serial Select" used to determine how the Arduino XBee module with the plate on which it is communicated. If the switch is placed in the "Micro" position, data received by the XBee module come without intermediaries to the microcontroller (RX signal) and the data sent from the microcontroller (TX signal) will be transmitted both to the module as to the possible computer it is connected via USB. However, that position ATmega328P may not be programmed via USB.

If the switch is set to the "USB" position, the XBee module will communicate directly with the Atmega16U2 chip (assuming you're using the plate UNO) completely ignoring the presence ATmega328P microcontroller. In this position, the XBee module will be able to communicate directly with a computer connected via USB to be configured and used independently of the other elements of the Arduino board. However, for the "USB" mode to work properly, you have to be careful to preset the ATmega328P microcontroller with an Arduino code with their duties "setup ()" and "loop ()" totally empty (in Chapter 4 It will be what this means).

This shield also incorporates a socket to place a microSD card, which can be used by programming library "SD" which comes standard in the Arduino language. To communicate with this card, the Arduino UNO uses the SPI protocol, using it as SS pin number 4, in addition to the pins 11, 12 and 13. This means that these four pins shield can not be used for else and therefore have 4 digital inputs / outputs less.

Arduino Wireless Proto Shield

This shield is exactly equal to the previous shield, but without the microSD socket.

Arduino WiFi Shield

This shield is intended for those who want to add to the Arduino UNO the ability to connect wirelessly to a TCP / IP network. HDG104 incorporates the chip manufacturer H & D Wireless, which includes a built-in antenna and can connect to Wi-Fi networks 802.11b 802.11g type. It also incorporates the 32UC3 ATmega chip, 32-bit microcontroller in this shield is factory preset to provide a complete IP stack (TCP and UDP).

The networks that can be connected can be opened or be protected by encryption type WEP or WPA2-Personal (can not connect to WPA2-Enterprise networks). In any case, you can only connect to a network if publicly broadcast its SSID (ie, if your SSID is not hidden). To manage this shield, use the library official program "WiFi" which comes standard in the Arduino language itself.
As with the rest of official shields once connected the shield plate on the UNO through the string of pins that fit perfectly above and below, to our circuits thereafter use the inputs and outputs offered by the pin-female This shield. These inputs and outputs have exactly the same layout and functionality as UNO plate. Even if necessary, it could seamlessly connect one second shield on top of the shield to keep adding functionality.

The process of loading our programs in the microcontroller board UNO coupled to shield WiFi unchanged compared to normally done with a plate ONE independent: we simply connect the board UNO to our computer via the USB cable to load the program, and once I have done, and we can continue forever, fueling more shield plate set via USB or disconnect the computer and plug it into the external power supply. The shield requires 5 V to operate, and this voltage is provided by the UNO plate by fitting the corresponding pin power between plate and shield.

This shield includes (as does the shield Ethernet or Wireless SD) A socket designed to place a microSD card. The idea is that this card will serve to store files and make them available through the network, using the programming library "SD" (which is the default in the Arduino language). Communication between the chip and the plate HDG104 ONE is set by pins
10, 11, 12 and 13 (via SPI) so these pins can not be used for other
purpose. Communication between the microSD card and the Arduino UNO is also set by the SPI protocol, but this time it uses as SS pin number 4 (instead of 10 used by the HDG104). In addition to this, the pin number 7 is also reserved for internal communication between the board and the WiFi shield. Therefore, we must bear in mind that unless we have 6-pin (4, 7, 10, 11,
12 and 13) because we can not use them as standard inputs or outputs.

In addition, the sharing of the channel between the HDG104 SPI chip and microSD card means that you can not use these two devices simultaneously. If we use in our programs both elements, we consider this to program your code correctly. If, however, we do not use one of the two devices have to be careful to explicitly disable it in our code; to clear the microSD card has been configured as an output pin 4 and writing a HIGH value and to deactivate -ALTO- HDG104 chip has to do the same but with the pin 10.

On the other hand, indicate that this shield also has its own button "reset" which resets both HDG104 chip as the Arduino itself, and also a number of interesting information LEDs to know: the "L9" (connected directly to digital pin # 9), the "LINK" (indicating that it has established a network connection), the "ERROR" (indicating whether there has been an error in communication) and "DATA" (indicating that Data is being transmitted / being received at that time).

Finally, we can see that this shield connector includes a USB mini-B type. This connector is not to program the Arduino board, but to perform a more advanced task that generally is not required and therefore not be discussed in this book: Update (DFU by protocol) ATmega32U3 firmware chip. A utility of this is to add directly to the 32U3 more complex network protocols without relying on the limited space provided by the ATmega328 on the Arduino board. In fact, even this may alter the firmware so that the WiFi shield could function as a standalone device without having an Arduino connected to it.

This shield also has a socket FTDI (to which you can connect an adapter or USB-serial cable) to interact directly with the chip
32U3 and diagnostic information. To do this, you must use a program
special computer that allows sending commands to the shield so that it interprets and executes. That is, a program that works as a "terminal number". This type of program will study in the next chapter. The list of the commands can be found here: http://arduino.cc/en/Hacking/WiFiShield32USerial.

Arduino Motor Shield

This shield incorporates the chip manufacturer STMicroelectronics L298P; "P" final simply indicates the encapsulation type you have, since for the same chip L298 there are other shapes and sizes, identified with other letters). This chip is designed to control components that contain inductors - "coils" - in their internal structure, such as relays, solenoids, DC motors -DC- or stepper motors - "steppers" -, among others.

Specifically, thanks to L298P chip, the Arduino Motor Shield allows us to control the speed and direction of rotation of up to two DC motors independently or these two magnitudes of a stepper motor. We may also perform various actions on the capabilities of the connected motors. To learn more about the different types of engines, their properties and applications, I recommend consulting the section dedicated to them in Chapter 5.

As the voltage necessary for the proper operation of engines usually exceed that provided by a simple USB cable, the shield will ever need to be powered by an external source. This can be either an AC / DC adapter to provide between 7 and 12 V DC output and connected to the 2.1 mm jack Arduino UNO, or a battery (preferably 9 V) connected by cables directly to the screw terminals labeled "Vin" and "GND" of the shield itself. In both cases (using the AC / DC adapter or battery), we would be feeding at once as both shield plate.

However, if the motors used require more than 9 V, we need separate power lines and shield plate so that each be fed separately: this is done by clearing the weld between the ends of the jumper labeled "Vin connect "on the back of the shield. Thus, the shield could continue to feed through its source connected to screw terminals, and the plate would feed its side or with the USB cable or the external source that was connected to jack 2.1 mm . However, keep in mind that the maximum voltage that support the screw terminals is 18 V.

This shield has two separate channels, labeled "A" and "B" in the form of screw terminals for connecting the engines there. Each channel individually can handle a separate DC motor, but can also be combined to operate between the two one stepper motor. If we are in the first case (that is, if we want to drive one or two DC motors), we can use certain pin-female input / output for controlling and monitoring the DC motor connected to each channel. This implies that those pin-female can not use for anything else, so that we will have less input / output general purpose. Specifically, these pins function-specialized female is:

Pines Pines Function	Channel A	Channel B
Direction of movement	D12	D13
Speed (PWM)	D3	D11
Brake	D9	D8
Current Sense	A0	A1

From the above table we can deduce how the operation of the shield. After connecting the cable pair to provide each motor to terminals (+) and (-) of the respective screw terminals (channel A or channel B), to control its direction of movement must send a signal of HIGH value or LOW

(Depending on the direction of rotation) to the corresponding address pins. To control the rotational speed must vary the values of the PWM signal corresponding pins. Finally, the brake pin HIGH if they have the value associated stop short DC motors (instead of leaving them gradually slow as would occur if the power is cut).

Can know the value of the intensity of current passing through the DC motor by reading the value of the mark in the corresponding pins in the table above, since the received signal is proportional to said intensity. Each channel can receive a current of 2 amps maximum, which maximum signal to a potential of 3.3 V corresponds

If you do not need the functionality of brake or current sensing and need more pins to our project, you can disable this functionality by clearing the weld between the ends of the respective jumpers clearly labeled on the back of the shield.

This shield also has a very interesting series of connectors: 2 white type connectors "TinkerKit" 3-pin to plug in two analog inputs (connected internally to pin A2 and A3); 2 orange connectors "TinkerKit" type 3-pin plug to two analog outputs (connected internally to the PWM outputs of the D5 and D6 pins) and 2 white type connectors "TinkerKit" 4 pin, one for input I2C / TWI and one for output I2C / TWI.

TinkerKit (and other)

"TinkerKit" (http://www.tinkerkit.com) is a prefabricated set of sensors and actuators that share a consistent and common form of coupling to any compatible circuit so that building projects as simple as connecting and disconnecting TinkerKit modules as if they were pieces of a puzzle. In this way, you can quickly launch interactive environments without welding or use even a breadboard. TinkerKit All components are mounted on a support generally orange and all have the same cable irrespective 3 (or 4) pin, which allows data transmissions at distances up to 5 meters. Among the components TinkerKit we can find buttons, LEDs of different colors, rotary and linear potentiometers, joysticks, accelerometers, motion sensors, light, infrared, temperature, transistors, actuators, compass, GPS, touch systems ...

For our Arduino UNO can recognize and work with components TinkerKit in general it is best to use the so-called "Sensor Shield," a shield prepared so that we can plug in there TinkerKit a centralized way the components you want. However, once the TinkerKit modules connected to the "Sensor shield" and this to an Arduino board, to manage the entire system is also necessary to use a special programming library called "TinkerKit" downloadable TinkerKit page (since comes not incorporated in the Arduino language) default. Using this library you can manipulate in your code each TinkerKit modules independently, thanks to a series of special instructions that facilitate their control.

There are other systems components that pursue the same philosophy of "plug and play", but each adopts its own connectors and can not be compatible. Alternatives to TinkerKit we can find for example the "GROVE Starter Kit" (product code ELB152D2P) Seeedstudio distributor. This product incorporates several GROVE modules, among which we find light sensors, temperature, water, gas, motion, distance, sound, electricity, touch sensors, accelerometers, compasses, liquid crystal displays, watches RTC, receiver / transmitter Bluetooth, GPS ... well kind potentiometers, joysticks, buttons, LEDs, buzzers, speakers, etc. It also includes the so-called "Grove Base Shield" which is simply a plug to an Arduino board connectors GROVE type where you can plug different modules of this type shield. GROVE modules to control, unlike with TinkerKit do not need to use any specific library: Arduino with language as it is already enough.

Seeedstudio kit also markets other modules (with the same philosophy but perhaps not as complete) called "Electronic Bricks Kit" (product code ELB138E1P), which includes the shield "Arduino Sensor Shield". Another "Sensor Shield" separate but compatible with the "Electronic Bricks" is distributed by Yourduino.

Arduino Proto Shield

This shield can easily design and implement our personal circuits. Basically what it does it is provide a work area which can be welded (using both technical -SMT- Surface mounting, as the technique of the through holes THT, in English, of "through-hole technology" -) the different components mails need to set up our project. This Thus, we can have a very compact form a complete whole circuit formed by plate, shield and connected components, all in one.

The THT technology is perhaps the easiest way to solder a component to a printed circuit board. For this, components are required to distribute metal pins several millimeters in length and the PCB available drilled holes: the technique simply involves crossing with each pin hole of the plate and then welding each of these pins to the underside of the PCB (optionally cutting, once the welding millimeters pin that can protrude below). However, with the SMT components but they do not have small metal tabs pins which are welded to a small special sockets present on the outer layer of the plates. The SMT technique, compared with THT, allows components to be much smaller, cheaper and can be used on both sides of the plates (thus allowing a much higher density components), but is generally more difficult to soldering for electronics hobbyists.

We clarify here because in this book project where no soldering required components is performed. While doing household welds (especially THT type) is not overly complicated and can make very good material and relatively cheap tools smooth results, we must recognize that some experience and expertise in this area is required. Therefore, and since this text continues to be an introduction to the world of electronics, it has not seen fit to investigate this matter, which may belong to a step immediately above knowledge.

Even if not be welded any component (as in our case), this shield also can be useful because we can permanently placed in an area specially reserved for it a small breadboard 1.8 x 1.4 inches (purchased separately) . So we can quickly plug all the circuit elements and ensure that this work properly in a very compact and safe way.

This shield is supplied with standard 5 V and GND pins of the Arduino and distributed along two rows eleven holes in the shield clearly labeled as "5 V" and "GND". From these two rows you can easily feed any component (including DIP sockets) being welded to the plate.

Other features of this shield is that it has a reset button and a ICSP connector (connected by pins 10, 11, 12 and 13 to the plate
Arduino UNO) so that we do not use those located on the motherboard, and replicating the exact layout of all pins of E / S to Arduino UNO to work with them as usual.

WHAT ARE UNOFFICIAL SHIELDS?

There are a lot of shields designed and built by the community that offer solutions to specific needs that officials have no shields. In this sense, it is advisable to consult the web http://www.shieldlist.org, offering a centralized list of virtually all existing models shields, classified by manufacturer. There also find the basic specifications of each shield, details of their connections and in general all the information necessary to understand its operation.

An example of shield we can be useful in several different projects is generically called "proto shields". Basically, what we offer such shields is an area of prototyping circuits to implement a much more compact form if we did through a separate breadboard (in addition to any switch, LED or king potentiometers, depending on model). An example of this type of shield we have just seen: the official Arduino Proto Shield, but other similar manufactured by third parties are, for example, the "Protoshield Pro" Freetronics or "Prototyping Shield for Arduino" from DFRobot.

Unfortunately, many "proto shields" are distributed in kit form (ie, with unassembled parts), so it is necessary, once acquired, all welded parts for a functional shield. Examples of the latter are the "MakerShield Kit" Makershed, the "Proto Shield for Arduino Kit" Adafruit, the "Kit for Arduino Protoshield" of Seeedstudio or "Arduino Protoshield Kit" Sparkfun.

An alternative to the "proto shields" above are generically called "screw shields", which replaced the pin-female terminals nut by 3.5 mm, ideal for securing cables in a much more stable form by screwdrivers. Examples are the "Terminal Shield for Arduino" of Freetronics, the "ProtoScrewShield" Sparkfun (in kit form), the "Pro- ScrewShield" Adafruit (in kit form), the "Arduino Proto Screw Shield" from IteadStudio or "Power ScrewShield" of Snootlab.

Another type of different shields of the above but you must know that we can also be of great help in many projects are those that allow themselves put on some kind of battery so that a solid and compact structure of supply is achieved more Arduino, making the energy-independent set of our computer. An example of this type of shield is the "Lithium BackPack" Liquidware, which supports placement (by JST connector) of a Li-Ion battery of various compatible types (2200 mAh, 1000 mAh or 660 mAh), all capable to feed our Arduino; also you have the option to act as a battery charger, either through food received by the Arduino board or directly from the outside through a USB mini-B connector already included.

Other similar to the previous shield is "Lipower Shield" SparkFun. This shield is also capable of performing two functions can be used as a battery charger (in this case, type LiPo 3.7 V and preferably 500 mAh) and also as a power source Arduino. The internal circuitry of the shield is responsible for protecting the battery charges, shocks or excessive currents, and to convert 3.7 V offered by it in the 5 V required to operate Arduino, so just worry about us we connect LiPo battery type JST socket built into the shield and go. If you want to recharge the battery, we can do it through the power received by the Arduino board itself or directly from the outside through a USB mini-B connector is included. This shield also allows you to monitor the state of charge of the battery using a specific downloadable Arduino code page of this product.

Anyway, it is impossible to mention here all the shields that may be outstanding for a reason: they are simply too many and very different. Many of them we will know as we need to use in the different existing projects throughout the book, but still remain in the inkwell many that we can be useful to know. I recommend, therefore, visit the website mentioned above and / or consult the extensive catalog of the manufacturers and designers of major shields, as Sparkfun, Adafruit, Seeedstudio, Iteadstudio, Freetronics, Olimex or DFRobot, among others listed in Appendix A.

ARDUINO SOFTWARE

What is an IDE?

A program is a set of specific instructions, sorted and grouped properly and unambiguously that aims to achieve a particular result. When we say that a microcontroller is "soft", we are saying that can record in memory permanently (until regrabemos again if necessary) the program we want to run this microcontroller. If we do not introduce any program in the microcontroller memory, it does not know what to do.

The acronym comes from IDE Integrated Development Environment, which translated into our language means Integrated Development Environment. This is simply a way to call the set of software tools that allows programmers to develop (ie, basically writing and testing) their own programs comfortably. For Arduino, we need an IDE that allows us to write and edit our program (also called "sketch" in the world of Arduino), allowing us to check that we have not made any mistakes and also allow us, when we are sure that the sketch is correct, record it in the memory of the microcontroller Arduino so that this becomes thereafter in the autonomous executor of the program.

To start developing our own sketches (or try one we have at hand) we install in our computer IDE that provides the Arduino project. To do this, follow any of the steps shown below, according to each particular case.

INSTALLATION Arduino Ubuntu

The easiest way to install Arduino IDE in Ubuntu is by Using its "Software Center" (or any other manager of equivalent packages such as Synaptic or apt-get / aptitude). The package you need is called "Arduino" and will be available if we have activated the "Universe" repository. If we use the terminal command to install the Arduino, so we could write: sudo apt-get -y install arduino

Interestingly Ubuntu divides the environment Arduino programming in two different packages: the so-called "Arduino" which includes the files that make up a graphical interface for IDE and called "arduino-core" which includes the essential tools for compiling, programming and recording sketches. The latter are all doing important work and are invoked using the graphical interface IDE (which in this sense does a mere intermediary between them and the user), but also can be executed directly by the shell. Installing the "Arduino" package will be installed automatically "arduino-core" package without having to specify it (it is in fact what we do with the above command), but the reverse is not true: if we work through a terminal without using the IDE itself could install "arduino-core" package independently without installing the "Arduino" package.

Fedora

The easiest way to install the Arduino in Fedora is using any of the available package managers (PackageKit, Yum, etc.). The package you need is called "Arduino" and is available in the official repository of Fedora. If we use the terminal command to install the Arduino should write: sudo yum install -y Arduino.

As was the case with Ubuntu, Fedora divides the Arduino programming environment in two packages: "Arduino" and "arduino-core". It also offers a third package called "Arduino-doc" (automatically installed by the above command), which contains all official documentation and examples of programming Arduino. In Ubuntu this information is included in the "arduino-core" package.

Any Linux system

To install the Arduino in any Linux distribution, http://arduino.cc/en/Main/Software must go to the official download page Arduino. There, under the "Downloads" section, there are two different links to download the version of the Arduino IDE for Linux. Both links point to an archive in "tgz" hosted at the official store of Arduino software (located on http://code.google.com/p/arduino) format, but a bond corresponds to the IDE version 32 and the other bits to 64 bits. It is necessary, therefore, first know what type of operating system we have installed (32-bit or 64-bit) and then unburden the "tgz" corresponding package.

To find out quickly if your Linux system is 32 bits or 64 bits can open a terminal and type the following command: getconf LONG_BIT output we get as a result of executing the above command will be "32" if our system is 32 bits and "64" if our system is 64 bits.

Once you know this, we can choose the "tgz" appropriate and, once downloaded, unzip the package. When we get inside the unzipped folder to obtain not find any installer or anything (just a structure of files and subfolders that we should not change anything) because in fact, our IDE is ready to be used from this moment: as we just click on the executable file (actually a shell script with execute permissions) called "Arduino" and, after confirming the "Run" will appear in the popup, see the IDE finally before us.

We can also run the IDE through the shell; for that we put ourselves through the cd command within the folder where the shell script and write ./arduino (important not to forget the dot and bar).

Obviously, we can move and store in the place we want our hard disk the folder containing the files that make up the IDE, as long as no we alter anything inside. Surely we want to create a shortcut to the executable wherever we deem appropriate (for example, on the desktop)

Actually this whole process just described in this section also could have also followed if our Linux distribution Ubuntu or Fedora outside. In fact, there may be occasions where we are interested more follow this "manual" procedure using the respective repositories: one of these occasions could be the appearance on the official website of a newer version of the IDE that has not yet been packaged or introduced into repositories of our favorite distribution.

Dependencies

Unfortunately, it is possible to try to run the IDE as described in this section, we get different errors due to not having previously installed on the system packages required to run the "Arduino" script without problems (ie, not having installed "dependencies" Arduino IDE). If we use a package manager to install the IDE either from your distribution repositories (such as "apt-get" Ubuntu or "yum" Fedora) will not have this problem because it is precisely one of the tasks of a package manager It is to detect whether the required dependencies are already installed (and if they are not, proceed with the automatic installation).

To resolve possible errors of lack of dependencies, we install them in some way before trying again to run the Arduino IDE. The easiest way is to do it through our favorite package manager, but what dependencies are? Here, we show reference list (not exhaustive!) With the names of the packages for Ubuntu and Fedora of the basic premises of the Arduino IDE, along with a brief explanation of their usefulness:

UBUNTU PACKAGE: openjdk-7-jre

FEDORA PACKAGE: java-1.7.0-openjdk

UTILITY: set of utilities and libraries that allow the execution of

applications written in the Java programming language. Precisely code IDE (unlike the basic command-line tools), is developed with this language, so installing this package is essential for correct operation on any operating system. His official website is: http://openjdk.java.net

UBUNTU PACKAGE: librxtx-java

FEDORA PACKAGE: rxtx

UTILITY: library written in Java that allows applications to communicate written in the same language (in our case, the Arduino IDE) with external devices using serial communication (in our case, the Arduino boards). In short, it is the piece of software that allows the IDE to communicate via the USB cable to the board that we connected. His official website is: http://rxtx.qbang.org.

UBUNTU PACKAGE: gcc-avr
FEDORA PACKAGE: avr-gcc and avr-gcc-c ++
UTILITY: collection of utilities that transform a source code written in the programming language C (or C ++) in a file with the appropriate binary format (specifying the format "hex") to be recorded when desired in memory of a Atmel AVR microcontroller. That is, a set of programs that transform our sketch in an understandable and executable file for this type of microcontrollers. Technically speaking, "gcc-avr" is a cross compiler (ie, a compiler that generates executable programs to a different platform from the platform where -Arduino- -PC- installed). The concept of compiler discussed below. His official website is http://gcc.gnu.org.

The package includes Ubuntu crossed for AVR C language and C ++ but Fedora are separated into two different packages compilers. Possibly surprised speak the languages C and C ++, but has already been previously commented that the language Arduino no longer a simpler subset based on these two languages (very similar to each other, in fact), so really, when Arduino sketches write our language, we are doing it without knowing basically a C / C ++ language simplified. Therefore, the basic tools compile are those of these two languages.

UBUNTU PACKAGE: avr-libc
FEDORA PACKAGE: avr-libc
UTILITY: A set of libraries needed to "gcc-avr" can make
compilations correctly. Technically, libraries are the standard C language for AVR platform. His official website is http://www.nongnu.org/avr-libc

Complementing "gcc-avr" (which is automatically installed as a dependency) must also have in the system the set of utilities generically called "binutils" (the corresponding package in Ubuntu is called "binutils-avr" and Fedora "AVR binutils "), which perform assembly tasks and linking libraries" avr-libc "and other libraries necessary for proper compilation. His official website is: http://www.gnu.org/software/binutils

UBUNTU PACKAGE: avrdude

FEDORA PACKAGE: avrdude

UTILITY: command line utility used to charge from a PC

files compiled by "gcc-avr" (with the help of "avr-libc" and "AVR binutils") in memory of type AVR microcontrollers (using the "bootloader" which incorporates, either directly through an ISP programmer). His official website is: http://www.nongnu.org/avrdude. In previous models Arduino boards to current rather than being used as a dependency AVRDUDE came another similar utility called Uisp (http://www.nongnu.org/uisp). Bossa (http://www.shumatech.com/web/products/bossa): In the Arduino Due model used a different utility.

We must clarify that all previous units, there are a few that actually we never download "by hand" because are included in the official Arduino package (but have been named for completeness). The reason for this inclusion is because versions of programs incorporated into the IDE are slightly different from the "standard" versions available in the repositories, and do not recommend using the latter because the IDE will not function. Specifically, they are the "rxtx" package (modified to support the plates using the ATmega16U2 chip instead of FTDI chip as Arduino UNO), the "avrdude" package (modified to ensure correct behavior auto-restart plates before loading the program, without which this load fail) and the "avr-gcc" package. The source code for these modified versions is available in http://github.com/arduino/Arduino

User permissions

In addition to installing all these units, we still make one last step before you can run the Arduino IDE: make sure that our system user has the necessary permissions for using the IDE, this can communicate through the USB cable with the plate . If we do this, only the IDE running as user "root" we can load programs into the microcontroller, which is not recommended.

To achieve this, Ubuntu just have to be careful to add our user to predefined users group "dialout". This pre-defined user group has already established the correct permissions to establish a serial communication with Arduino boards that connect to the computer. If our user does not belong to this group (itself usually belong default), can be solved by opening a command prompt and typing: sudo usermod -G -a myuser dialout where "myuser" I have to change the name of our User system, where the parameter "-G dialout" specifies the group to which you want to add and where the -a switch indicates that only want to add a new user group without removing it from the other possible groups they might belong above.

For Fedora, we see that our user also belongs to the group "dialout" and again, to another preset group, the "lock". Once relevant to our user groups, either Ubuntu or Fedora assigned, we restart the session for the changes to take effect.

On the recognition and use USB devices in Linux-ACM

When an Arduino UNO board is connected to a computer running a Linux system, several things happen:

1. The plate (more specifically ATmega16U2 chip) is recognized as a device type "USB ACM". This is easy to check this by running the command dmesg and observing its output.

Regardless of the operating system used, the type ACM USB devices ("Abstract Control Modem") have the characteristic of being able to establish with the computer serial communication simple and straightforward via USB. They are actually a subset of a larger set of devices, those belonging to the CDC ("Communications Device Class") class, which in turn are a small set of all possible chips that meet the USB standard.

2. Automatically generates the file / dev / ttyACM #, where "#" will be a different number depending on the number of devices of the same type that have connected at that time (we assume from now on that plaque is recognized as / dev / ttyACM0). Similarly, when the board is disconnected from the computer, the file / dev / ttyACM0 automatically disappears. It is easy to verify the existence of this file watching dmesg or also running the command ls -l

/ Dev / ttyACM0.

In order to recognize the peripheral-type "USB ACM" and create the device file / dev / ttyACM0 relevant, the Linux kernel uses a module (which is always the most widespread distributions is integrated by default) called "cdc_acm". Thanks to this module is working, we can establish a serial communication between our Linux system and device represented by

/ Dev / ttyACM0.

If an Ubuntu or Fedora distribution command output "ls" specified in section 2, we can see that the owner of this file is the user "root" and the owner of the file group is the "dialout" group observed , which have read and write permissions on the file. This is why we need to include our system user in the "dialout" group: to have read and write permissions on that file so you can send and receive data through its associated device. We may also assign read and write permissions to the group "other" (with the "chmod" command for example) but is not a very elegant solution.

The reason that Fedora also need our user belongs to the group "lock" is because in this distribution, this group owns and has all privileges in the / run / lock / lockdev, directory used by the kernel to block a possible simultaneous access by multiple users plate system. In short: to ensure that our users can communicate with it, you need full permissions (read, write and access -ejecución-) on that directory. In Ubuntu this step is not necessary because the directory in question already has full permissions default to any user of the system (thus a more permissive factory settings).

It is worth knowing that any chip that complies with the USB standard need a product identifier ("id product" PID) and a manufacturer ID ("Vendor ID" VID) officers own unique order to be marketed . The VID codes are sold to different hardware manufacturers in the world by the international USB-IF (http://www.usb.org) corporation, responsible for the standardization and specification of the USB protocol, and PID codes are elected by each manufacturers on their own. For example, has the VID FTDI 0403 and the Arduino Team, meanwhile, he has also bought its own VID (2341) and uses the Arduino UNO.

Any Linux system (from source)

The source code Arduino programming environment is available for download http://github.com/arduino/Arduino. However, this
option is only recommended for advanced users; to install the IDE on a Linux system quickly and conveniently are advised to follow the steps described in the previous sections. If, even so, it is preferred to compile the source code IDE to better suit the characteristics of our computer (or even to modify personally some internal functionality) in the following link you can find step by step instructions for successful process: http://code.google.com/p/arduino/wiki/BuildingArduino.

On the other hand, if you want to participate in developing the IDE, suggesting improvements or reporting errors in http://code.google.com/p/arduino/issues/list we can see the list of problems in the IDE code unresolved and management; http://arduino.cc/pipermail/developers_arduino.cc and you can access the mailing list of developers, where consultations on the functioning of the code are made and ideas to optimize and enrich suggested, besides discussing the design of different plates and official Arduino hardware.

Windows

To install Arduino IDE on Windows 8, we go to their official website downloads: http://arduino.cc/en/Main/Software. There appear under the "Downloads" section, a link to download the version of the IDE for Windows, which is simply a compressed zip file. Therefore, the first thing we have to do is once downloaded decompress. When we get inside the unzipped folder to obtain not find any installer or anything (just a structure of files and subfolders that we should not change anything) because our IDE is ready to be used from this moment: we just click on the executable file "arduino.exe" and appear before us.

Obviously, we can move and store in the place we want our hard disk the folder containing the files that make up the IDE, as long as no we alter anything inside. Surely we want to create a shortcut to the executable wherever we deem appropriate.

However, we have not finished the process of installing the IDE because Windows 8 does not include the required standard "drivers" to recognize the Arduino board when this is connected to our computer via USB. So in order to start working with our board we must first install these drivers in our system. Assuming that our plate is the Arduino UNO model, the steps to take are:

1. Connect the USB Arduino UNO through our computer.

2. Access the "Device Manager" of your Windows 8. For this, we place ourselves inside the box "Desktop" and once there, move the mouse cursor in the upper right corner of the screen. a side menu icons will appear; there must select the "Settings", and then, on the side panel icon newly emerged, we must choose the "Control Panel" option. A window that shows different categories will open; what interests us is "System and Security". Clicking on it we can finally display the link that leads to the "Device Manager".

3. From the list that shows the "Device Manager", we see an "unknown device". Select it, we click the right mouse button and choose the "Update Driver Software" option.

4. Chose "Search for driver software on your computer" and navigate to the folder containing the files "inf", which include the necessary information so that Windows can finally install the correct driver (there is a file ".inf" by each model of Arduino). The folder where these files are called "Drivers" and is located inside the unzipped folder containing the files of the IDE. Eye, not to be confused with the subfolder "FTDI USB Drivers".

5. Are we to ignore the warning from Windows to the absence of digital signatures in Arduino drivers. After this step, the installation process should be completed without a problem. To verify that everything went well, we should see the list displayed by the "Device Manager" a new device within the category "Ports (COM and LPT)" called "Arduino UNO (COMxx)", where "xx" will be a number (typically be 3 or 4, but can be another). This number will be important later remember to use our card correctly.

Mac OS X

To install Arduino IDE on Mac OS X we go to their official website downloads: http://arduino.cc/en/Main/Software. There will appear under the "Downloads" section, a link to download the version of the IDE for Mac OS X, t is simply a compressed zip file. Therefore, the first thing we have to do is once downloaded decompress. Then we get a "dmg" file; by double clicking on it will start the standard installation process any native application on Macintosh systems. The process is almost entirely automatic: the only thing we do is, in the window that opens us to the beginning of the installation, drag the icon onto the Applications folder Arduino for the IDE is installed there (although in fact we could choose install elsewhere). And that's it: if all goes well, the launcher bar the icon of the newly installed Arduino will appear.

First contact with the IDE

Once installed the Arduino IDE (in the system it), nothing more start it you will see a window similar to this (the catch is made using the IDE version 1.0.2, which is current as of publication of this book) :

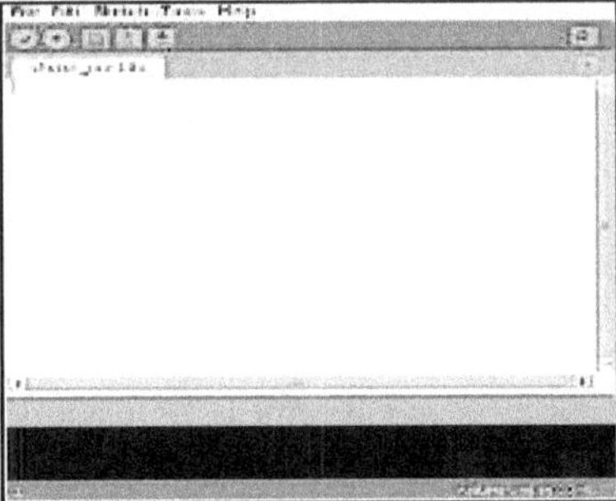

We can see that the IDE window is divided into five major areas. From top to bottom they are: the menu bar, the toolbar, the code editor itself, bar and console messages, and the status bar.

IDE area where work longer will the code editor, since that is where we will write our sketches. Another area often use the button bar is composed of the following elements:

Verify: This button does two things: check that there is no error in the code of our sketch, and if the code is correct, then compiled. This is the first button you have to press every time we want to try any changes we make in our sketch. (If you do not know what "compile", do not worry: now you just have to know that it is an essential step).

Upload: this button we pressed the button immediately after "Verify". Its function is internally invoke the command "avrdude" for load into memory microcontroller Arduino sketch recently verified and compiled. "Avrdude" is able to perform its task because clicking this button the bootloader of the microcontroller (the "auto-reset") is automatically activated so rarely be necessary to use the "traditional" method to activate the bootloader by physical press of the reset button next to the plate.

New: Creates a new empty sketch.

Open: displays a menu of all the sketches available to open. We can open both our own sketches as many sketches ready to test sample, categorized in the menu. These sketches are very useful for learning; in fact, in this text we will use enough of them, because they are public domain.

Save: saves the code of our sketch to a file, which will have the ".ino" extension. We can save these files anywhere we want, but the Arduino IDE offers a specific folder for it, the folder "Sketchbook" located within the personal user folder on your system and automatically generated the first time the IDE runs. In fact, within this folder "sketchbook" a different subfolder for each project within which the corresponding sketches will be saved will be created; thus sketches of different projects are not mixed together.

Serial Monitor Opens the "monitor series." We talk him right away.

We can also see that just below the button "Serial Monitor" have a drop-down button from which we can open new tabs. Have multiple tabs open at once can be useful when we have such a long code we need to divide it into parts to work more comfortably. This is because all new open tabs are part of the same project as the first original tab (and therefore all of them written in code is global, one) but the particular content of each of the tabs physically stored in a different file, allowing easier handling. In these cases, when a project consists of several source files, opening one with programming environment, it detects the existence of the other files included in the project and automatically displayed on the corresponding tabs. The most common is to use separate tabs for defining functions, constants or global variables (concepts to be studied in the next chapter).

Other trivial actions you can perform with the dropdown button are close the current tab, rename the current tab (resulting in renaming the sketch included in it), move to the next or previous tab, or move to a specific tab.

The bar menu offers five main entries: "File", "Edit", "Sketch", "Tools" and "Help". We clarify that possibly these names and in fact, all elements of the SDI input and see translated to the language set by default in your operating system, but to avoid possible inconsistencies in this book continue naming by its name Anglo, which is the original. The entries in the menu bar showing additional actions complement those that can be performed with the button bar. Interestingly, the menus are context sensitive, meaning that only the relevant elements at a given time (as we are doing) will be available, and those that do not appear disabled. Some of the actions you can perform using the menus are:

Menu "File" and offer standard actions as creating a new sketch, open an existing one, save it, close it, close the IDE itself, etc., can also see other interesting actions. For example, thanks to the "Examples" entry you can access the sketches of example that come standard with the IDE and thanks to the "Sketchbook" entry you can access our own sketches saved in different subfolders inside the folder " sketchbook. " Other actions to be considered are eg loading the sketch in the microcontroller memory (equivalent to "Upload" button seen above), the loading sketch in memory using an external ISP programmer (previously selected from the list of compatible ISP programmers available on the menu "Tools" -> "Programmer") and printing the code sketch by printer (can also configure the format of the page). Finally, thanks to the "Preferences" entry, we can open a pop-up box that offers the possibility to set some preferences IDE, such as changing the location of the "sketchbook" folder, the language of the IDE, the size of the font, activating the automatic updates of the IDE, the level of detail in the messages displayed during the compilation process and / or load of sketches, etc.

There are many more preferences that do not appear in the popup. If we change them, we must do it "by hand" by editing the appropriate value in the configuration file "preferences.txt" which is simply a text file containing a list of data pairs <-> value fairly self-explanatory. Depending on your operating system, the file "preferences.txt" will be located in a different folder, folder we have already shown in the pop himself. Anyway, this file has to be modified when the IDE is not running, otherwise, the changes will be overwritten by the environment itself when closed.

"Edit" menu: besides offering standard actions such as undo and redo, cut, copy, and paste text, select all the text or find and replace text, we can see other interesting actions. For example, thanks to the "Copy for forum" input you can copy the code of our sketch to the clipboard in our system in a way that is especially suitable for paste thereupon directly in the official forum of Arduino (and thus can receive assistance community). Thanks to the "Copy as HTML" input you can copy the code to the clipboard our sketch of our system in a way that is particularly suitable for pasting into generic web pages. Other actions to be considered are for example to comment / uncomment the portion of text that we have selected, or apply or remove indentation (more on the importance of these two possibilities in the next chapter).

Menu "Sketch": In this menu provides the action to check / compile our sketch (equivalent to "Verify" button seen earlier), open the folder where it is saved the ".ino" file being edited

at this time, adding a new tab in a new code file to our sketch and import libraries.

Menu "Tools": In this menu different variety of tools, including the ability to auto-format the code to make it more readable (eg, bleeding the lines of text contained within the braces are offered
{And}), the ability to save a copy of all the sketches of the current project in zip format, the ability to open the serial monitor (equivalent to the button "Serial Monitor"), etc. Other more advanced tools are for example the "Programmer" entry already mentioned or "Burn bootloader" input useful when you want to record a new bootloader in the micro plate. Require special mention entries "Board" and "Serial", which will be discussed more fully in the following paragraphs.

Menu "Help": from this menu you can access various sections of the official Arduino website containing different articles, tutorials and examples help. No Internet needed to consult these sections as this documentation is downloaded together with the IDE itself, so that their access is done locally (ie, "offline"). Consultation is strongly recommended.

The bar and console messages inform at the time of the compilation of possible mistakes in writing our sketch, besides indicating the real-time status of various processes, such as file recording "ino" to disk hard, sketch compilation, loading the microcontroller, etc. It is also interesting to note that every time a successful compilation takes place, the size that occupy the sketch inside the Flash memory of the microcontroller displayed in the console messages. Is information is very valuable, because if the size of our sketch were close to the maximum allowed size (in the case of the Arduino UNO remember that is 32KB), could know and then conveniently modify the code of our sketch to decrease in size and do not overdo the limit.

The status bar on the left simply shows the line number of the current sketch where the cursor is located, and to his left the kind of Arduino and the computer serial port currently used right now.

The "Serial Monitor" and other series terminals

The "Serial Monitor" is a window of IDE that allows us, our computer to send and receive text data to the Arduino using the USB cable (more accurately, using a serial connection). To send data, simply type the desired text in the box that appears at the top and click on the "Send" button (or press Enter) text. Although obviously not help this shipment if the board is not programmed with a sketch to be able to get this data and process them. On the other hand, data received from the plate will be displayed in the center section of "Serial Monitor".

It is important to choose the right combo box through the bottom of the "Serial monitor" the same transmission rate (in bits / s, also called "baud") that is specified in the sketch running on the board, because if no, the characters will not be recognized correctly transferred and the communication does not make sense.

It is also important to remember that in MacOSX and Linux systems Arduino UNO is autoresetea (ie restarts to run the sketch from the beginning) every time we open the "Serial Monitor" and connect with her. This plate with Leonardo does not occur.

Not only through the "Serial Monitor" we can communicate via a serial connection (via USB) to the plate. You can use any other program to send and receive data through such connections. These programs are often called "serial terminal". In the Ubuntu and Fedora (and most Linux distributions) we find several, such as "GTKterm", "cutecom", "picocom" or "minicom", among others. In Windows you can use the "telnet" command that comes by default in a standard installation of the system or third-party programs such as "putty" (http://www.putty.org) or "Terminalbpp" (https: // sites. google.com/site/terminalbpp), among others. In Mac OS X is a good application "CoolTerm" (http://freeware.the-meiers.org).

In any of the above programs you do not need any special configuration to work properly with our plate: only required to specify the speed (in bits / s) at which data transmission and the serial port to be used will be made to establish communication (for example, / Dev / ttyACM0 on Linux, Windows COM3, etc .; the possible values discussed in the next section) and you're done.

SETTING AND CHECKING THE CORRECT OPERATION IDE

Once you have connected via the USB cable newfound our plate to our computer, the first thing we should see is that the LED labeled "ON" lights up and stays on continuously always so. After checking this, we can start the Arduino programming environment. Once opened however, before you can start writing a single line we must ensure that the IDE "knows" two basic things:

1. The type of Arduino connected to the computer at this time (UNO, Leonardo, Mega, etc.)

2. The serial port of your computer that is to be used to communicate via USB with it.

For the former, we should just go to the "Tools" menu -> "Boards" and select from the list that the board appears to be working. There fixed plate models appearing several times as if they can come up with different models of microcontroller. For the latter, we must go to "Tools" menu -> "Serial port" and choose the appropriate serial port (can be real or virtual). Depending on the operating system used in that menu options appear. The most common are:

Linux: the device files appear (usually
/ Dev / ttyACM #).

Windows systems: a list of ports COM #, where "#" is a number appears. Typically this number will be 3 or higher, since the COM1 and COM2 ports are typically usually reserved for real hardware serial ports (not simulated via USB). If there are multiple COM ports and do not know what is the corresponding to the Arduino, as simple is disconnected and reopen the menu: the entrance is gone will we seek; connecting plate can again select smoothly.

MacOS X systems: shows the device files (usually
/dev/tty.usbmodem###). It is possible that at the time of connecting a
UNO or Mega board a popup informing you that it has detected a new network device is displayed. If this happens, just click on "Network Preferences ..." and when the new window appears, simply click on "Apply". The plate is shown as "Not configured", but work properly and so you can leave the system preferences.

Both configurations (and port plate) will be needed to specify them only once while not modify (ie, while not use other plate, for example).

Now we only need to check that everything works correctly, test running some sketch. As we write none yet on our own, we will use a sketch of coming as an example. Specifically, the sketch called "Blink". To open it, we must go to "File" menu -

> "Examples" -> "01.Basics" -> "Blink". Now we are not interested in studying the code shown: we simply click on the "Verify" button, then click the "Upload" button. We always follow this pattern with all the sketches: once want to try how they work, we must first press "Verify" (to compile and verify the code) and then "Upload" (for loading into the microcontroller). Just after you press the last button, it happens that:

1. First the LED flashes very quickly labeled "L" indicating that the board is reset and, therefore, it is running the bootloader

2. Then the board LEDs labeled "RX" and "TX" quickly flash several times, indicating that the sketch is coming to the plate and thus being received by the bootloader and loaded into the flash memory microcontroller. In the IDE we observe the process status reporting messages, and its successful completion.

3. Finally, once finished charging, what should happen is that the LED
labeled "L" will begin to blink periodically, and for always. If so, congratulations !: happens our board works perfectly and can be programmed by our computer smoothly.

ARDUINO beyond language: the language C / C ++

Throughout the preceding paragraphs we have repeated several times the word "compilation" but we have not explained their meaning until now. "Compile" means converting a code written in the IDE (using the language Arduino, and therefore understandable easily by humans) in the really executable program by the microcontroller, which is nothing more than a huge set of bits (ie 1s and 0s) only understandable for him.

That is, we write in the IDE instructions using simple Arduino language and subsequently by compiling, we transform the sketch into "digestible" instructions for the microcontroller (in what comes to be called "machine code" or "binary code "). This machine code is essentially nothing more than a set of electrical pulses (1s Come in and current-
'It's current-0s), which is all that really know how to process electronic circuits. Thus, a resulting machine code (fictional) compilation of an Arduino sketch anyone could look like this: 10010111010101011011 ...

Clearly it is impossible to write a program directly in machine code; therefore compilers exist. But in addition, these tools offer another advantage: you have to know that a valid machine code for a microcontroller is not for another, due to different internal electronic construction. Therefore, the same program would have to encode different machine codes for different models of microcontroller. However, if you have specific for each of these models compilers from the same source code we can get different machine using in each case the respective compiler. This is a great advantage because we can not rewrite already functional code or mistakes in our developments, as well as providing great flexibility and scalability to our sketches.

Since Arduino boards incorporate AVR microcontrollers architecture (except Due, which is ARM, but we will consider it), it stands to reason that the compiler included in the official programming environment Arduino is a specific code to generate binary compatible with such chips. And so it is. However, the tool (called "gcc-avr") does not compile code written in Arduino language into binary code AVR (specifically in "hex" format), it compiles written in C / C ++ language into binary code AVR code. Why? Because really, Arduino language is not a language in the strict sense: it is simply a set of instructions C and C ++ "camouflaged", designed to simplify the development of programs for AVR microcontrollers. That is, when we're writing our sketch in "Arduino language" without knowing we are actually programming in a simplified version of C / C ++ language. If you want to know in detail all the internal transformation of the sketch originally written in Arduino language version in C ++ language and its subsequent compilation "gcc-avr" in http://code.google.com/p/arduino / wiki / BuildProcess is available the necessary information.

The C language and its "relative" C ++ are two of the most important and widespread in the world for several reasons languages, because they are powerful languages and yet light and flexible, because they have a vast ecosystem of libraries that provide it with functionality that other languages do not offer, because written in these languages and compiled programs are extremely efficient and fast, and because there compilers for virtually any type of hardware (so that today we can see a multitude of software written with these languages running on a variety Machine).

However, perhaps a person that begins in the world of programming languages C / C ++ are not too friendly. In other words, they are relatively difficult languages to learn and master. So there Arduino language: for a person with little knowledge of application development can quickly write a functional Arduino sketch without having to learn a whole programming language complete but complex as C or C ++. That is, Arduino language makes "mask" of the C / C ++ language, hiding lot of extraneous detail for the enthusiast of electronic projects and facilitating the use of the Arduino platform altogether. The "price" to pay for winning this ease of programming sketches is that we have only a subset of all the functionality that may offer the C / C ++ language. Fortunately, for most projects you need not go beyond what is already offered by the language and its official Arduino IDE, and of course, all the examples in this book are programmed using only the language features offered exclusively Arduino .

IDEs ALTERNATIVE TO OFFICIAL

There are people all over the world that the official Arduino will just not convincing, for several reasons: lack of advanced features (such as auto-complete sentences, for example) or Java dependency (which implies to have installed in our computer more packages than you actually need to compile and load our sketches) are few. Another reason is that many developers are already familiar with a particular IDE and want to continue using it for their developments with Arduino. The consequence of this is that there is currently a variety of "alternative" IDEs that provide more features than the original IDE or just changing the way people work. Then some of these IDEs are named (not an exhaustive list!) If the reader wants to look at them a feature that has no official IDE:

CodeBlocks (http://www.codeblocks.org) is a free IDE and platform for developing applications written in C and C ++ language. However, you can write programs directly on Arduino language (and also load on the plate) using a modified version of this software, called "CodeBlocks Arduino Edition", downloadable from the direction http://www.arduinodev.com/codeblocks .

Gnoduino (http://gnome.eu.org): free IDE that tries to mimic in appearance and functionality to the officer, but avoids its reliance on Java, as it is written in another language, Python. This makes it extremely light and fast. However, it only works in Linux, and more specifically within the GNOME desktop since its dependencies.

Codebender (http://www.codebender.cc): This IDE is actually a web application, so it works entirely in an "online" within the browser, without installing anything. It includes a full text editor, a compiler and charger sketches, all without leaving the browser. In addition, previous free registration, can store "in the cloud" made the code set. This project is free, so you can download and Install the necessary to mount a Codebender own software platform.

Visualmicro (http://visualmicro.codeplex.com) actually is a supplement ("plug-in") to use the Arduino language within the programming environment Visual Studio (http://www.microsoft.com/ VisualStudio) of Microsoft. Unfortunately it does not work in this environment Express version (free), so it has to buy a higher edition, such as Professional.
EmbedXcode (http://embedxcode.weebly.com) actually is a supplement ("plug-in") programming environment from Apple officially called XCode (https://developer.apple.com/xcode) . It permits an IDE "all in one" by which you can program in a unified way different platforms such as Arduino, chipKIT, Maple, MSP430 or Wiring (although each using its own programming language).

Anyway, if you do not want to use any of the supported development environments before, we still have the ability to write Arduino code with absolutely any other IDE we want. To do this, we have installed our favorite IDE both as the official Arduino environment, and follow the steps below. In this way we can use all the features for editing code providing favorite IDE and use only the official environment to compile and load.

1. Run the Arduino IDE and open the ".ino" file you wish to edit.

2. Go to the IDE preferences box and enable the "Use external editor". Automatically, the code editor is gray, indicating that it is disabled.

3. Run the IDE we want, and open the same ".ino" file. Perform the necessary editing.

4. Save the changes in the editor used. It is important that the ".ino" recorded file keeps the same name as the ".ino" file originally opened in the Arduino IDE, otherwise this will not know.

5. Use the buttons "Verify" and "Upload" the Arduino IDE as usual, when deemed appropriate.

On the other hand, it also noted the existence of a kind of Arduino development environment somewhat "special" because they are focused on visual programming sketches. Ie instead of using written instructions in an abstract language, codes are constructed from the graph "coupling" of different colored blocks that represent shares and control structures. Its aim is to enable anyone without any programming experience (eg, children) initiation into the world of microcontrollers, thus killing two birds with one stone: the introduction to programming and introduction to electronics. Highlights include: Scratch for Arduino -S4A- (http://seaside.citilab.eu/scratch/arduino) is a plug-in visual programming environment Scratch (http://scratch.mit.edu) allowing use this to schedule and interact with Arduino boards.

Modkit Micro (http://www.modk.it): visual development environment much like Scratch. It can be used in more architectures besides Arduino boards, and has the distinction of being able to work "online" within the browser. You can also download the installable environment, but only for Windows and MacOSX. Another feature is that if you want, you can work with the internal code beyond visual blocks, allowing more direct customization of sketches.

Minibloq (http://blog.minibloq.org): same as above, a visual environment is ready to be used in more architectures Arduino alone, but only works on Windows programming.

Ardublock (http://blog.ardublock.com): visual development environment for Arduino programmed in Java (and therefore cross-platform).

Finally emphasize, though not a development environment itself, the Ino (http://inotool.org) tool. It is an executable program directly from the command interpreter that allows you to compile and load Arduino sketches. These sketches, though, must have been written and saved previously using any text editor. It also includes an option equivalent to "Serial Monitor". Your configuration details (such as plate model or serial port to use at all times) can be specified as command parameters or within a specific configuration file. It is recommended to consult the relevant documentation on their official website to learn its use. Unfortunately, it does not work on Windows systems.

ARDUINO LANGUAGE

MY FIRST SKETCH ARDUINO

Arduino connects to your computer and runs the official IDE. Select (if not already) the appropriate type of plate (in the Tools-> Board menu) and the USB port used (in the Tools-> Serial Port menu).

Example 4.1: Create a new sketch with the following contents.

```
/*Declaración e inicialización de una variable
  global llamada " mivariable" */
int mivariable=555;
void setup() {
        Serial.begin(9600);
}
void loop() { Serial.println(mivariable);
        mivariable=mivariable+1;
```

Click the "Verify" button, then click the "Upload" button. There you should see no error messages on the console. Open now the "Serial Monitor" and see that there are appearing in real time many numbers one after another, starting with 555 and followed by the 556, 557, 558, 559, etc., increasing steadily. Why? What does this text (code) that we have introduced in the microcontroller memory of the plate?

GENERAL STRUCTURE OF A SKETCH

A program designed to run on an Arduino (a "sketch")
always it consists of three sections:

The section global variable declarations: located directly at the beginning of the sketch.

The section called "void setup ()" defined by opening and closing keys.

The section called "void loop ()" defined by opening and closing keys.

The first section of the sketch (which does not have any delimiter symbol start or end) is reserved for writing, as its name suggests, the various declarations of variables we need. In a subsequent section largely explain what all this means.

Inside the other two sections (ie, within their keys) must write the instructions we want to execute on our plate, taking into account:

Written instructions in the "void setup ()" are executed only once, when turning (or reset) the Arduino board.

Written instructions in the "void loop ()" are executed just after the section "void setup ()" countless times until the plate is off (or reset). That is, the content of "void loop ()" will run from the 1st instruction to the last, to then re-run from the 1st instruction to the last, to then run from the 1st instruction to the last, and on and again.

Therefore, the written instructions in the "void setup ()" normally used to perform some initial presets and instructions inside "void loop ()" are, in fact, the program itself is running continuously.

In the case of example 4.1, we see that in the declarations of global variables is a single line (int myvar = 555;) that within "void setup ()" one instruction (Serial.begin (9600) runs ;) and that within "void loop ()" continuous and repeated execution is performed (until the power is interrupted plate) of two statements one after another:
Serial.println (myvar); and myvar myvar = + 1 ;. On the meaning and syntax of these lines discussed below.

On the case, tabs and semicolons

It should be clarified and little details that we take into account when writing our sketches to spare us a lot of headaches. For example, it is necessary to know the Arduino language is "case-sensitive". This means that it is completely different type a letter in case insensitive. In other words: for the Arduino language "hello" and "HELLO" are two different words. This has an important implication: it is not the same type eg "Serial.begin (9600);" that "Serial.begin (9600);". In the first case would be properly written instruction, but in the second, when compiling the code the IDE to complain because he "serial" (with "s" lower case) makes no sense. So you have to watch a lot with respect this distinction in the code we write.

Another detail: the tabulations of the instructions contained in the sections "void setup ()" and "void loop ()" the sketch of example 4.1 are not at all necessary for compiling the sketch takes place successfully. They are simply a way to write code in an orderly, clear and comfortable for the programmer, making it easier for reading and writing code and maintain a certain structure at the time of writing. In the following examples in this book you are better off watching their usefulness.

Another detail: all instructions (also including variable declarations) end with a semicolon. It is essential always add this sign to avoid compilation errors, because the compiler needs to locate to detect the end of each instruction written in the sketch. If you forget, an error text may be obvious ("missing semicolon") or not show. If the error text is very dark or without logic, it is good idea to check that the cause is not the lack of a semicolon in just before the line marked by the compiler as the cause of the problem lines.

REVIEWS

The first line of the sketch of example 4.1 contains a comment (Specifically, are the first two lines: from the symbols / * to the symbols * /). A "comment" is a written text interleaved with the code that sketch used to report on how the code works to the person who is reading some point. That is, the text comments are helpful to humans that explains the associated code and help to understand and remember its function. Comments are completely ignored and discarded by the compiler, so never not part of binary code that executes the microcontroller (so do not take up space on your memory).

Comments may appear within the code in different ways:

Comments composed of an entire line (or part of it) to add we write two slashes (//) at the beginning of each line you want to comment out. We can also discuss only part of the line, if we write the bars at another point than the beginning of this; only in this way we will be discussing what appears behind bars until the end of the line, but this does not.

Comments consist of a block of several consecutive lines: to add we write a slash followed by an asterisk (/ *) at the beginning of the block of text you want to convert to comment, and an asterisk followed by a slash (* /) at the end of that block. All characters and lines located between these two marks start and end are automatically treated as comments. Keep in mind, moreover, that unilineal comments can be written within multiline comment, but one within another multiline no. This is the kind of comment written in the sketch of example 4.1.

A fairly common practice in programming is to comment at some point one (or more) parts of the code. Thus, the "erased" those parts (ie, are ignored and therefore not compile or run) without actually deleting them. Typically, this is done to locate possible errors in the code by observing the behavior of the program with those particular lines commented. Throughout the examples in this book you will see its usefulness.

VARIABLES

The first line of the sketch of example 4.1 is to declare a global variable of type "int" called "myvar" and initialize it with the value of 555. Let us explain this.
A variable is an element of our sketch that acts as a small "drawer" (identified by a name chosen by us) storing a particular content. That content (what is called the value of the variable) may be modified at any time of the execution of the sketch: hence the name "variable". The importance of the variables is huge, as all sketches make use of them to accommodate the values they need to function.

The value of a variable may have been obtained in various ways: literally may have been assigned (as in Example 4.1, where just start explicitly assigned to the variable called "myvar" value 555), but the data can also be obtained by a sensor, or the result of a calculation, etc. Initially from whatever comes, that value will always be changed at any later time of the execution of the sketch, if you wish.

Declaration and initialization of a variable

Before you can start using any variable in our sketch; however, we must first create it. The fact of creating a variable is often called him "declaring a variable". Arduino language, when you declare a variable must also specify its type. The type of a variable you choose as data type (integer, decimal numbers, string, etc.) that we want to store in that variable. That is, each variable can only store values of a certain type, so we must decide in his statement what type of variable is what interests us more depending on the type of data we anticipate store. Assign a value to a variable that is of a different type than expected for this error causes a sketch.

Possible types of the Arduino language will detail in the following paragraphs, but you can know that the general syntax of a statement is always well written online: variableName tipoVariable ;. In the case of wanting to declare several variables of the same type rather than write a separate statement for each, you can declare them all on one line, so generically: tipoVariable nombreVariable1, nombreVariable2 ;.

Optionally, while the variable is declared, you can set an initial value: this is called "initialize a variable". Initialize a variable when you declare it is not mandatory, but strongly recommended. 4.1 In the example, declare a variable called "myvar" type "int" (which is a data type from the various existing designed to store integers) and additionally also initialized with the initial value of 555. Hence, and it can be deduced that the general syntax of an initialization is always well written online: tipoVariable variableName = valorInicialVariable ;.

One might ask why it is convenient to use initialized variables instead of writing the value directly where necessary. For example, in the case of the 4.1 code example, we assigned the value 555 to "myvar" when we could have written directly into the Serial.println () statement like this: Serial.println (555) ;. The main reason is because we work with variables greatly facilitates the understanding and maintenance of our programs: If this value is written in many different lines of our code and must be changed, using a variable initialized to that value should only change variable initialization and automatically every time where that variable appear within our code would this new value; if we wrote directly that value in each line, would Irlo changing line by line, with the consequent loss of time and the possibility of errors.

On the other hand, when declaring variables, you should give our variables descriptive names to make our code more readable sketch. For example, names like "sensorDistancia" or "botonEncendido" help you better understand what these variables represent. To name a variable you can use any word you want, whenever this is no longer a reserved word of the Arduino language (such as a command name, etc.) and do not start with a digit.

Assigning values to a variable

What happens if a variable is declared but not initialized? You will have a default value (usually without interest for us) until you assign a different value at some point in the execution of our sketch. The general syntax to assign a new value to a variable (either because it has not been initialized, or especially because they want to overwrite a previous value with a new one) is: variableName = newValue ;.

For example, if the variable called "myvar" and is of type integer, it could be written within any section of our sketch a line as myvar = 567; to assign the new numerical value of 567. Important notice that the assignment line reads "left to right": the "newValue" value is assigned to the variable "variableName".

The assignments of values to variables can be varied: direct values are not always as in Example 4.1. A rather common case is assigned a value that depends on the value of another variable. For example, if we assume a variable named "y" and another called "x", we could write the following:
y = x + 10 ;. The above statement is to be read thus: the value it currently has the variable "x" is added 10 and the result is assigned to the variable "y" (that is, if for example "x" has a value of 5 , "and" will have a value of 15).

We can even find assignments y = y + 1; (As indeed it does the sketch example 4.1). The key here is to understand that the "=" is not the mathematical equality (which we will study) but the assignment. Therefore, a line like the one above it does is add a drive to the current value of the variable "y" (understood to be a numeric type) to then assign this again resulting new value to the variable "and" overwriting which had previously. That is, if initially "and" vale 45 (for example), after executing the assignment y = y + 1; worth 46.

Scope of a variable

Important other variables concept is a variable field. In this sense, a variable can be "global" or "local". It is an area or another depends on where in our sketch declare the variable:

For a variable to be global it has to declare the beginning of our sketch; that is, before (and outside) the sections "void setup ()" and "void loop ()". In fact, speaking of the structure of a sketch we mentioned the existence of this section global variable declarations. A global variable is one that can be used and manipulated from anywhere in the sketch. That is, all instructions of our program, no matter in which section spelling ("void setup ()", "void loop ()" or others that may exist) can view and also change the value of that variable.

For a variable is local has been declared within any section of our sketch (ie, in "void setup ()" or "void loop ()" or others that may exist). A local variable is one that can only be used and manipulated by the written instructions in the same section where stated. This type of complex variables is useful is to ensure that only one section has access to their own sketches and long variables, as this avoids errors when a section inadvertently modify variables used by another section.
Possible types of variable

The types of variables that Arduino language supports are:

The type "boolean" variables of this type can only have two values: true or false. They are used to store a status between these two possible, and so make the skech react as detected in them either. For example, Boolean variables can be used to check whether data has been received one (true) sensor or not (false) to check whether any actuator is available (true) or not (false) to check if the value of another different variable meets a particular condition such as being greater than a specific number (true) or not (false). The value stored in a boolean variable always occupies one byte of memory.

To explicitly assign a variable of type "boolean" true value, you can use the special word "true" (without quotation marks) or the value "1" (without quotation marks), and assign the value false can use the special word "false" (without quotation marks) or "0" (without quotes). That is, if in our example sketch of the variable "myvar" would have been of type "boolean" instead of "int", to assign the value of "true" should have written a line similar to myvariable boolean = true; or boolean myvar = 1; (Actually, a boolean variable with a different value of either 0 is interpreted as having a certain value: it has to be specifically set to 1).

The type "char" means the value that can have a variable of this type is always one character (letter, digit, punctuation ...). If we want to store a string (ie, a word or a phrase) the "char" type does not help us, we use other explained later.

To explicitly assign a variable of type "char" a certain value (ie character), we must be careful to write that character in single quotes. Therefore, if the sketch of example 4.1 the "myvar" would have been varying type "char" instead of "int", to assign the value of the letter A should have written a line similar to myvariable char = 'A ';.

In fact, the characters are stored internally as numbers and electronic devices are unable to work with "letters" directly: the have to "translate" always first numbers then they can be stored and processed. To find out which internal number corresponds to a particular character, and
vice versa, the Arduino board uses the call ASCII table, which is a simple list of equivalences that associates each character with a certain number.

The fact that the characters for Arduino actually be recognized as numbers allows it possible to perform arithmetic operations with these characters (or rather, with its corresponding numerical value in the ASCII table). For example, if we perform the operation 'A' + 1 would get the value
66, as in the ASCII table the numerical value of the character 'A' it is 65. In fact, even
We could initialize a variable "char" by assigning a numerical value instead of the corresponding character (ie: char line myvar = 'A'; we could write as char myvar = 65; and both are equivalent).

Each variable of type character occupies 8 bits (one byte) of memory to store its value. This means that there are 28 = 256 different possible values for a variable of this type (it is easy to see that if combinations of 0s and 1s that can be obtained with 8 positions are counted). Since the values of type "char" they are actually numbers, values that can store a variable of this type are within the range from -128 to 127 number.
The type "byte": the value that can have a variable of this type is always an integer between 0 and 255. As the variables of type "char", the type "byte" uses a byte (8 bits) to store value and therefore have the same number of different possible number combinations (256), but unlike those, the values of a variable "byte" can not be negative.

The type "int": the value that can have a variable of this type is an integer between -32768 (-215) and 32767 (215-1) by using 2 bytes (16 bits) of memory for storage. This is true for all boards except for Arduino Due: in this model of plate type "int" uses 4 bytes, and therefore, its value may be within a wider range, especially between -2,147,483,648 (-231) and 2,147,483,647 (231 -1).

The type "word" type variables "word" in the Arduino Due occupy 4 bytes to store its value. Therefore, they have the same number of different possible number combinations that variables "int", but unlike these, the values of a variable "word" can not be negative. Based on plates of type AVR microcontroller same thing happens: variables of type "int" and "word" occupy the same memory space (although in this case, however, are only 2 bytes) but the values of the second They can not be negative. It is easy to see the value that can have a variable "word" on all boards except the Arduino Due is an integer between 0 and 65535 (216-1).

The type "short" means the value that can have a variable of this type for all models of plate (whether based on microcontrollers such AVR ARM -the majority or type -the Due-) is an integer between - 32768 (-215) and
32767 (215-1), by using 2 bytes (16 bits) of memory
stored. Here, the "int" type "short" and to plates
AVR family are equivalent but the Arduino Due to the type "short"
It is the only one using 16 bits.

The type "long" means the value that can have a variable of this type for all models of plate (whether based on microcontrollers such AVR or ARM type) is an integer between 2,147,483,648 and 2,147,483,647 thanks they use 4 bytes (32 bits) of memory for storage. In this sense, the types "long" and "int" for the ARM family plates are equivalent.
The type "unsigned long": the value that can have a variable of this type for all models of plate (whether based on microcontrollers such AVR or ARM) is an integer from 0 to 4,294,967,295 (232-1) . As variables of type "long", the type "unsigned long" use 4 bytes (32 bits) to store its value, and therefore have the same number of different possible number combinations (232), but unlike of those, the values of a variable "unsigned long" can not be negative (as its name already indicates). Here, the types "unsigned long" and "word" to ARM family plates are equivalent.
In the preceding paragraphs commented, referring to the numerical variables, they all have a range of valid values, and therefore assign a value outside this can have unexpected consequences. This must be taken into account to learn properly choose the type of numeric variable we need. For integer variables, which you might think it is to always use the type "unsigned long", which allowed the highest range. But this is not a good idea, because every variable of this type occupies four times more than a variable "byte" and if (for example) know beforehand that the values stored will not be greater than 100, it would be a complete loss of memory the variables used "unsigned long". And the microcontroller memory is one of the most precious and scarce for missing out by watching it a poor choice of resource types.

In the event that a value exceeds the valid range (both "above" and "below"), what will happen is that "give back" and the value will continue at the other end. That is, if we have a variable of type "byte" (for example) whose value is currently 255 and will add 1, its new value will then
0 (if it will add 2, then would be worth 1, and so). The same happens if the lower limit is exceeded: if we have a variable of type "byte" (for example) whose current value is 0 and subtract 1, its new value is then 255 (if you subtract two, it would then
254, and so on). In fact, this can see if you run the sample sketch
4.1 and wait long enough: when the values of "myvar" variable (of type

"Int", remember) reaches its upper limit (32767), see how the next value automatically returns to the principle of rank (namely -32768) to continue increasing again steadily to a peak and return to dropping back to a minimum, and so on. This phenomenon is called "overflow".

There are more data types supported by the Arduino language in addition to the already mentioned:

Type "float" means the value that can have a variable of this type is a decimal number. The "float" values can range from any number
-3.4028235 · 1038 to 1038 · the number 3.4028235. Because of its great range of possible values, decimal numbers are often used to approximate continuous analog values. However, only having 6 or 7 digits total precision. That is, the "float" values are not accurate, and can produce unexpected results, for example, 6.0 / 3.0 does not give exactly 2.0.

Another drawback of the values of type "float" is the mathematical calculation with them is much slower than integer values, so using values "float" in parts of our sketch that need to be run at high speed should be avoided.

Decimal numbers have to write in our sketch using the Anglo-Saxon notation (ie, using the decimal point instead of a comma). If you wish, you can also use scientific notation (ie, the number
0.0234 -equivalent to 2.34 · 10-2 - we could write such as 2.34e-2).

The type "double" is a synonym for the type exactly equivalent "float" and therefore provides no increase in accuracy with respect to this (unlike what happens in other languages, where "double" does yield twice precision). Both a variable of type "double" type as a "float" occupy four bytes of memory.

The type "array": this type of data does not exist as such. So there are arrays of variables of type "boolean" type variables arrays "int" type variables arrays "float", etc. In short: arrays of variables of any type mentioned heretofore. An array (also called "vector") is a collection of variables of a particular type that are all one and the same name, but can be distinguished from each other by a number as an index. That is, instead of having different variables for example type "char" - each independent of the other (varChar1, VARCHAR2, varChar3 ...) we have a unique array that groups all under one name (eg , varchar), and to allow each variable can be handled separately by that within each array is identified by an index number, written in brackets (varchar [0], VARCHAR [1], VARCHAR [2] ...). The arrays are used to gain clarity and simplicity in the code and facilitates programming.

We can create an array -declarar- (whether in the area of global declarations or within any particular section), in the following ways:

varInt int [6]; 6 declares an array elements (ie individual variables) uninitialized none.

varInt int [] = {2,5,6,7}; Declares an array without specifying the number of elements. However, they are assigned (in braces, separated by commas) values directly to individual elements, so that the compiler is able to deduct the total number of array elements (in the example on the right, four).

varInt int [8] = {2,5,6,7}; Declares an array of 8 elements and initializes some of them (the first four), leaving the rest uninitialized. Logically, if more elements are initialized to allowing the array size (for example, if nine values are assigned to an array of 8 elements), an error would occur.

char varchar [6] = "hello";
char varchar [6] = {'h', 'o', 'l', 'a'};
varchar char [] = "hello";

The first form declares and initializes an array of six elements of type "char". As arrays of type "char" are strings reality of characters (ie words or sentences, "strings" in English) have the particularity to be initialized as shown in the first form: indicate directly the word or phrase written between double quotes. But they can also be declared as a "standard" array, which is like showing the second form. Observe in this case the difference in quotes: an individual character is always specified in single quotes, and the value of a string is always specified in double quotes. It is also possible, as shown in the third form, declare a string without specifying its size (as the compiler can deduce from the number of elements - ie, characters-initialized).

Keep in mind that the first value in the array has the index 0 and therefore, the last index value will equal the number of elements in the array minus one. Careful with this, because assigning values beyond the declared number of array elements is a mistake. In particular, if for example we have an array with 2 elements of type integer (ie, declared as follows: int varInt [2]), to assign a new value (eg 27) to its first element should write:
varInt [0] = 27; And to assign the second value (eg, 17), we would write
varInt [1] = 17; . Asignáramos but also a third value so varInt [2] =
53; , Making a mistake because we would be exceeding the expected end of the array (and therefore using a memory area not reserved, with unpredictable results).

On the other hand, besides giving an element of an array an explicit value as just mentioned in the previous paragraph, it is possible to assign an array element value you have at that moment another independent variable (preferably same type). For example, by varInt [4] = x line; we will be assigning the current value of a variable called "x" to the fifth element (the index starts with 0!) the array called "varInt". Conversely it is also possible to assign the value it has at that time the fifth element of the array "varInt" the independent variable "x", we simply run the line: x = varInt [4];

In the case of arrays of characters (the "chains" or "strings"), must take into account a very important feature: this kind of arrays should always be held with a number of elements one greater than the maximum number we anticipate saving character. That is, if you are going to store the word "hello" (four letters), the array must be declared at least 5 elements. This is because the latter element is always used to automatically store a special character (the character "zero" with ASCII code 0) used to mark the end of the chain. This brand is necessary for the compiler knows that the chain is over and do not try to continue reading more positions. If you do not know beforehand what is the length of the text is saved in a character array, we can declare this with a number of elements large enough for there to be elements without being assigned, knowing that so possibly we will be wasting memory microcontroller .

Finally, commenting that it is often convenient, when working with large amounts of text (for example, in a project with LCD screens), use arrays of strings. To declare an array of this kind, the special data type "char *" is used (note the trailing asterisk). An example of declaration and initialization of this type would be: char * varCadenas [] = {"Cad0", "str1", "str2", "Cad3"} ;.

Actually, the asterisk in the above statement indicates that we are actually declaring an array of "pointers" as to the Arduino language, strings are pointers. The "leaders" are elements of the Arduino language (from the C language in which it is based) very powerful yet

certainly complex. This book will not be discussed, because their potential uses are advanced and can confuse the reader that starts in programming: all you need to know is how the string arrays are declared and no more. The good news is that once declared the array of strings, we can work with it (assigning values to its elements by consulting, etc.) as any other type of array without notice at all that we are using pointers.

Changing data type (numeric)

You can never change the type of a variable is declared if a particular type, will remain such throughout the entire sketch. But what we can do is change "on the fly" at a particular time the type of the value it contains. This is called "casting", and can be useful when you want to use that value in calculations that require a different type of original, or even when you want to assign the value of a variable from one type to another variable of a different type. To convert a value-the kind that sea- another, we can use any of the following instructions:

char (): write in () the value, or the name of the variable that you want to convert in contiene- type "char"
byte (): write in () the value, or the name of the variable that you want to convert in contiene- type "byte"
int (): write in () the value, or the name of the variable that you want to convert in contiene- type "int"
word (): write in () the value, or the name of the variable that you want to convert in contiene- type "word"
long (): write in () the value, or the name of the variable that you want to convert in contiene- type "long"
float (): write in () the value, or the name of the variable that you want to convert in contiene- type "float"

Example 4.2: Consider the following illustrative code:

```
float variablefloat=3.4; byte
variablebyte=126;        void
setup() {
```

```
		Serial.begin(9600);

		Serial.println(byte(variablefloat));

		Serial.println(int(variablefloat));

		Serial.println(word(variablefloat));

		Serial.println(long(variablefloat));


		Serial.println(char(variablebyte));

		Serial.println(float(variablebyte));
}
void loop() {
   //No se ejecuta nada aquí
}
```

You can see, once the previous sketch, all conversions of value float to integer values (regardless of whether they are "byte", "int", "word" or "long") truncate the result executed: had a 3.4 have to have a 3. The difference is that the bytes occupied memory 3. We can also see that when converting a numerical value to type "char", the corresponding character in the ASCII table shows. Finally, the conversion of an integer to a decimal value causes can work precisely with decimals thereafter.

Example 4.3: A specific situation where we need to control the types of variables are correct it is in the mathematical calculation. For example, if the following code is executed, you can see that the result is 2 when it should be 2.5.

```
float   resultado;   int

numerador=5;         int

denominador=2;  void

setup() {

		Serial.begin(9600);

		resultado=numerador/denominador;

		Serial.println(resultado);

}
void loop() {}
```

Why is this happening? For both the "numerator" variables as "denominator" they are whole, and therefore the result is always full, even though we may observe in a variable of type "float". This is because the result of a mathematical calculation where different integer types is always of integer type with increased use of memory and not lose information-sign-for but never decimal involved. Only if involved in calculating a value of type decimal, then the result is DECIMAL (although it remains our responsibility to keep that value in a variable of type decimal that there is no truncation).

To solve the above code, we must make a "casting" to "float" from one of the two elements of the division (or both if you want) for the operation to be performed using these new types, and get a result as "retipeado" ready to be assigned to the variable of type decimal. That is, we have to replace result = numerator / denominator; by, for example, result = float (numerator) / denominator ;.

Example 4.4: Another problem with the change of data types you can see in the following code:

```
int    numero=100;
long      resultado;
void setup() {
          Serial.begin(9600);
          resultado=numero*1000;
          Serial.println(resultado);
}
void loop() {}
```

When you run the above code we will see the "Serial Monitor" a number that is not the correct result of multiplying 100 by 1000. Why, if the variable "result" is a "long" and therefore can store a number of this magnitude? Because the value that is assigned is the result of multiplying the value of "number" (type "int") by the number 1000 (which, if not explicitly specified, is of type "int" also because we know all written number is literally always type "int"). That is, two values of type "int" are multiplied, and the result therefore yes exceeds the accepted range such data. The variable "result" is of type "long" does not influence the number obtained by multiplying two numbers "int" cease to be wrong, because the calculation of the multiplication is performed before assigning the result to "result" . That is, when the value "result" is assigned, this has already suffered the overflow. To avoid this, the easiest thing would be to force any of the elements in the calculation is the same type as the variable "result" because, as we mentioned above, the data type of the resulting number is the same the operand data type with greater capacity. Therefore, if for example we make the variable "number" is "long", the multiplication result will be of type "long", which "overflow" will occur and eventually assigned correctly to the variable " result. " That is to say:

When you run the above code we will see the "Serial Monitor" a number that is not the correct result of multiplying 100 by 1000. Why, if the variable "result" is a "long" and therefore can store a number of this magnitude? Because the value that is assigned is the result of multiplying the value of "number" (type "int") by the number 1000 (which, if not explicitly specified, is of type "int" also because we know all written number is literally always type "int"). That is, two values of type "int" are multiplied, and the result therefore yes exceeds the accepted range such data. The variable "result" is of type "long" does not influence the number obtained by multiplying two numbers "int" cease to be wrong, because the calculation of the multiplication is performed before assigning the result to "result" . That is, when the value "result" is assigned, this has already suffered the overflow. To avoid this, the easiest thing would be to force any of the elements in the calculation is the same type as the variable "result" because, as we mentioned above, the data type of the resulting number is the same the operand data type with greater capacity. Therefore, if for example we make the variable "number" is "long", the multiplication result will be of type "long", which "overflow" will occur and eventually assigned correctly to the variable " result. " That is to say:

```
int numero=100;
long resultado;

void    setup()    {    Serial.begin(9600);
            resultado=long(numero)*1000;
            Serial.println(resultado);
}
void loop() {}
```

We know that when inside the Arduino code directly write numbers, it assumes by default that are of type "int". However, if after literal value add the letter "U", the default type is "word", if we add "L", its type will be "long" and if "UL" is added, its type is "unsigned long ". That is, a line number * result = 1000L; I had gotten the same effect as this line in the above code.

CONSTANT

You can declare a variable so that get their value (of any kind) always remain unchanged. That is, their value can not be changed because it is never marked as "read only". In fact, this type of variables is no longer called so for obvious reasons, but "constant". The constants can be used as any variable of the same type, but if you try to change its value, the compiler will throw an error.

To convert a variable (either global or local) constantly, all you have to do is precede the declaration of that variable with the const keyword. For example, to convert constantly a variable called "sensor" type "byte" simply has to declare as follows: const byte sensor; .

There is another way to declare constants in the Arduino language, which is using the special #define directive (inherited from the C language). However, the use of const is recommended for its flexibility and versatility.

PARAMETERS OF AN INSTRUCTION

Before you begin to learn and use different language instructions provided Arduino, we must be clear on one fundamental concept: the parameters of an instruction. Already it has noticed that in the example sketch
4.1, after the name of each instruction used (Serial.begin (), Serial.println ()
etc.) always some parentheses appear. These brackets may be empty but may also include within it a number / letter / word, or two, or three, etc. (If more than one value must be separated by commas). Each of these values is what is called a "parameter" and the number of them and its type (which does not have to be the same for all) will depend on each particular instruction.

The parameters are used to modify the behavior of instruction in some way. That is, the instructions do not have a single function parameters (his given task) and point: there is no possibility of change because they always do the same in the same way. When an instruction, however, has one or more parameters, it will also make its preset function, but the concrete is given by the value of each of its parameters, which modify some specific feature of the action to be performed.

For example: suppose you have a command called lalala () that has (that is "receiving", technically speaking) a parameter. Suppose if the parameter is 0 (ie, if lalala (0) is written;), which will make this instruction is to print out "Hello, friend"; if the parameter is 1 (ie, writing lalala (1);), which will be printed to the screen "How are you?" and if the parameter is 2 (ie, writing lalala (2);) print " Very good ", etc. Obviously, in the three possibilities lalala command () does essentially the same thing: to print out (for that is a command, but could be talking about three different commands), but depending on the value of its single parameter, the print action by A prompt is modified in some way. Summing: instructions would be as verbs (executed orders), and the parameters would be as adverbs (say how those orders are executed).

Note that in this hypothetical example, we are using a numeric parameter; in this case it would not give another type of value (such as a letter) because the instruction lalala () would not be ready to receive and give error. Each parameter must have a predefined data type.

Return value of an instruction

Another key concept related to the Arduino language instruction is to "return value". The instructions, in addition to receiving input parameters (if they do), and apart from homework they have to do, usually also return an output value (or "return"). An output value is a fact that we can get in our sketch as "tangible" results of instruction execution. The meaning of the returned value depends on each specific instruction: some are control (indicating whether the instruction has been executed well or badly), others are numerical results after the execution of a mathematical calculation, etc.

We know that every time you come to a line in our sketch where the name of the command appears with its possible parameters, is executed. We did not know it is that the return value of the execution automatically "replaced" within the code to the name of the instruction and made this substitution and continues to run the rest of the line. To understand why, suppose we have an instruction called lalala () that returns a certain value. If we want to use the value returned can:

Assign that value to a variable of the same type, so you can use it later. That is, if the variable unavariable call her for example, should write something like: unavariable = lalala () ;. Set in this case, as we have said, once the instruction executed, his name is "replaced" by its return value, and then this is assigned to unavariable.

Use that value (which we insist, "replaces" the name of the instruction where you are written when this is executed) directly into another instruction. For example, to see the value returned by lalala () we could run and send the "Serial Monitor" that value, all at once, like this: Serial.println (lalala ());

If you do not want to use at any time the return value, need not do anything special: just run the statement in the usual way and ready.

The serial communication with Arduino

We have already explained in previous chapters ATmega328P microcontroller has a receiver / transmitter UART TTL-series type that allows the Arduino UNO communicate with other devices (usually, our computer), so you can transfer data between them. The physical communication channel in these cases is usually the USB cable, but can also be digital pins 0 (RX) and 1 (TX) of the plate. If these two pins to communicate the plate with an external device is used, we have to concretely connect the TX pin of the plate with the RX pin of the device, the RX TX plate with the device and
share the earth plate with the earth of the device. Keep in mind that if these two pins for serial communication are used, they can not then be used as input / standard digital outputs.

Within our sketches we can make use of this receiver / transmitter UART TTL-microcontroller to send data to (or receive it) thanks to the Arduino language element named "Serial". Actually, "Serial" is what we call an "object" language. Objects are entities that represent specific elements of our sketch. The concept of object is abstract, but to understand it better, simply assume that they are "containers" that bring together different instructions with some relationship between them. For example, "Serial" object represents by itself a series communication established with the plate, and in our sketch we can use a set of instructions available within it that serve to manipulate this serial communication. If we use two objects of class "Serial" we could then handle two different serial connections, and each would be controlled by the instructions of the respective object.

Existing instructions within an object (not all Arduino language instructions belong to an object) are written following the syntax nombreObjeto.nombreInstruccion () ;. So the instructions used in the sketch example of the beginning of this chapter, belonging to the "Serial" object have names like Serial.begin () or Serial.println ().

Then we explain the syntax, operation and use of the instructions in the "Serial" object. We begin with an already known:

Serial.begin (): opens the serial channel for communication can start it. Therefore, implementation is essential before any transmission that channel. So normally it is usually written in the "void setup ()" section. In addition, through its only parameter -of type "long" and compulsory- specifies the speed in bits / s to the series of data transfer will occur. For communication with a computer, usually you use the value of 9600, but you can specify any other speed. What is important is that the value entered as a parameter matches that specified in the dropdown on the "Serial Monitor" Arduino IDE, or if no communication will not be well-timed and meaningless symbols are displayed. This instruction has no return value.

There is also the Serial.end () statement, which has no arguments and returns nothing, and is responsible for closing the channel number; in this way, serial communication is disabled and the RX and TX pins become available for general purpose input / output. To reopen the channel number again, it should be used again Serial.begin ().

The other instruction that we used in the previous example sketches is Serial.println (), which belongs to a set of instructions that allow you to send data from the microcontroller to the outside. Estudiémoslas whole.

Instructions for sending data from the plate to the outside

Serial.print () sent via a data channel number (specified as a parameter) from the microcontroller to the outside. This data can be of any type: character, string, integer, decimal number (default of two decimal places), etc. If the data is explicitly specified (instead of through a variable), remember that the characters have been written between single quotes and double-quoted strings.

In the event that the data sent is full, you can specify an optional second parameter that can be worth a predefined constant of the following: BIN, HEX or DEC. In the first case the binary representation of the number will be sent in the second, the hexadecimal representation, and the third, the decimal representation (the one used by default)

In the event that the data to send is decimal, you can also specify an optional second parameter to indicate the number of decimal places you want to use (by default there are two).

Its return value is data type "byte" is worth the number of bytes sent. In the case of strings, this return value matches the number of characters sent. The use or not in our sketch of this return value will depend on our needs.

The transmission of data by Serial.print () is asynchronous. That means that our sketch goes to the next instruction and continues to run without waiting for start sending data performed. If this behavior is not desired, you can add just after Serial.print () instruction Serial.flush (), which has no parameters and returns no return-value, instruction waits until the data transmission is complete for the continued implementation of the sketch.

Serial.println () does exactly the same as Serial.print (), but also automatically at the end of the data sent adds two additional characters: the carriage return (ASCII code No. 13) and newline (code ASCII No.
10). The consequence is that at the end of the execution of Serial.println () a line break occurs. It has the same parameters and return the same values that Serial.print ()

You can simulate the behavior of Serial.println () by Serial.print () if you manually add these characters. The carriage return character is represented by the symbol '\ r', and newline with '\ n'. So Serial.print ("hello \ r \ n"); It would be tantamount to Serial.println ("hello"); . Other non-printable ASCII character which can represent useful is the tab (using '\ t').

At this point, we have already seen the meaning of all the lines of our first sketch of this chapter, therefore, we should be able to understand their behavior. Let us analyze, then. If we remember the code, first we declared a global variable of type "int" and inicializábamos worth

555. Next, tearing his program execution opening the serial channel (at a rate of 9600 bits / s) so that the board could communicate with our computer. And finally, as in the "void loop ()", first we sent the current value of the variable to our computer (which is supposed to be the USB device connected via serial channel to open). Then we increased that value in one unit, for just resubmit that new value to the computer, and return to a unit increase in its value, and resend ... so infinitely until the plate stop receiving power. What we would see the "Serial Monitor" would be precisely those values (one in each different line because Serial.println () introduces an automatic line) that would increase one by one without stopping. At the time it reached the maximum value allowed by the data type of the variable (in the case of an "int" is 32767) it would follow for the minimum value (-32768) and goes on from there, an endless cycle.

It should also know the existence of the Serial.write instruction (), like Serial.print () but not the same:

Serial.write (): number sent via a data channel (specified as parameter) from the microcontroller to the outside. But unlike Serial.print (), the data to be sent can only occupy one byte. Therefore, it must be basically "char" or "byte". Actually, it is also capable of

transmitting character strings that are treated as a mere sequence of independent bytes one after another. However, other types of data, which occupy more than one byte inextricably (such as "int", "word", "float" ...) will not be sent correctly. Its return value is, as in Serial.print (), a data type "byte" is worth the number of bytes sent.

The grace of Serial.write () is that the data is always sent directly without interpreting. That is, it is sent as a byte (or series of bytes) as is without any format conversion. This does not happen with Serial.print (), which can be played with the binary, hexadecimal, etc. Therefore, this statement is intended for direct data transfer to another device without a preview on our part.

However, if we look at the "Serial Monitor" data sent by Serial.write (), we see that is (whether they are of type "char" and type "byte") as corresponding ASCII characters. This is because it is the proper "Serial Monitor" which performs real-time this "translation" ASCII. Thanks to this behavior, we can always see the character associated with the byte sent, which is what we usually want.

Example 4.5: The following code illustrates better the behavior just described:

```
char      cadena[]="hola";
byte bytesDevueltos; void
setup() {
          Serial.begin(9600);
          bytesDevueltos=Serial.write(cadena);
          Serial.println(bytesDevueltos);
}
void loop() {}
```

If you run the above code and observe the "Serial Monitor", we will see the value "hola4" appears. 4 shows the end value of the variable "bytesDevueltos" keeping it returned by Serial.write (chain) ;, confirming that have been sent by the serial channel 4 bytes, corresponding precisely to the string "hello".

There is another way to use Serial.write (), it is about to hit an array data type "byte". In that case, this command has two parameters: the name of the array and the number of elements (always starting with
First) you want to send. The latter value does not have to match the total number of array elements.

4.6 Example: The following code illustrates better the behavior just described:

```
//El array ha de ser de tipo " byte" (o " char" )
byte arraybytes[ ]={65,66,67,68};
void      setup()          {
        Serial.begin(9600);
        //Se envían solo los dos primeros elementos de ese array
        Serial.write(arraybytes,2);
}
void loop() {}
```

Anyway, byte transmission is asynchronous. That means that our sketch goes to the next instruction to continue running without waiting for start sending the / os byte / s made. If this behavior is not desired, you can add just after Serial.write () instruction Serial.flush ()
which has no parameters and returns no return-value, which
it will wait until the data transmission is complete for the continued implementation of the sketch.

Instructions for receiving data from the outside

So far we have seen instructions that allows the microcontroller to send data to its environment. But how to reverse it done? That is: how can send data to the microcontroller (for this pick them up and processed) coming from its environment, such as our computer?

From the "Serial Monitor" send data to the plate is very simple: there is only write what we want in the box shown here and press the "Send" button. However, if the sketch that is running on the board is not ready to receive and process this data, the transmission will not reach anywhere. Therefore we need in our sketches conveniently receive the data coming to the plate via serial communication. To do this, we have two basic instructions: Serial.available () and Serial.read ().

Serial.available (): returns the number of bytes available -caracteres- to be read from abroad through the serial channel (via USB or via pin TX / RX). These bytes have arrived to the microcontroller and remain temporarily stored in a small memory of 64 bytes TTL-UART has -called "buffer" chip - until they are processed by Serial.read () instruction. If no bytes to read, this command will return 0. No parameters.

Serial.read (): returns the first byte still unread which are stored in the input buffer TTL-UART chip. In doing so, it removes it from the buffer. To return (read) the next byte has been rerunning Serial.read (). And do so until they have read them all. When no more bytes available, Serial.read () will return -1. No parameters.

Example 4.7: Consider a basic code of these new instructions:

```
byte byteRecibido = 0;
void setup() { Serial.begin(9600);
}
void loop() {
```

```
if (Serial.available() > 0) { byteRecibido =
        Serial.read();        Serial.write("Byte
        recibido: ");
        Serial.write(byteRecibido);
}
```

In the above code we have introduced an element of language that we have not yet seen: the conditional "if". We will explain in detail in the corresponding section of this chapter, but suffice it to know that an "if" look if written in brackets condition is true or not: if it is, the written instructions within their keys are executed, and if no, no. The condition is Serial.available see that ()> 0, so that what is being tested is whether or not data stored in the input buffer TTL-UART chip. If this condition is true, inside the "if" is executed, basically what it does is to read the first byte that is available in the buffer (removing it from there) and send it to "Serial Monitor". As all this occurs within the "void loop ()", then back again to its beginning again to check if there are still data stored in the input buffer. If this continues, the next available byte is read and sent back to the "Serial Monitor". And we will check again if there is still data in the buffer, in which case the next byte will be read. And so until they have read all available bytes. At that time, the condition of "if" becomes false (because Serial.available () return 0) and therefore "void loop ()" will not execute anything. But as each repetition of "void loop ()" will continue

It checking whether there is buffered data, when to do again, the condition of "if" again be true and therefore they again begin to read the bytes available there, one by one.

A very important detail of the code above is that, as you can see, the value returned by Serial.read () is stored in a variable of type "byte". This implies that the value is stored in numeric format. That is, if Serial.read () receives eg the value "a", which is stored in a variable of type "byte" is the corresponding numerical value in the ASCII table (in this case, 97); equally, if Serial.read () receives eg the value "1", actually what is stored in a variable of type "byte" it is the numerical value 49, and so on. And this is independent of how the data is displayed by the "Serial Monitor". However, if the type of the variable used to store the value returned by Serial.read () is "char" (remember that is the other data that occupies 1 byte in memory), what happens is that "a" It will be read as the character "a", "1" as the character "1", etc. The consequence of this is that we must first think what use we will give you our sketch the return value (number or character?) Then choose the data type of the variable that will save it.

There are other instructions in addition to Serial.read () that read data input buffer TTL-UART chip more specific forms, which we can come in handy in certain circumstances:

Serial.peek (): returns the first byte still unread which are stored in the input buffer. However, unlike Serial.read (), the byte read buffer is not cleared, so the next time you run Serial.peek (), or once Serial.read () - will be re-read same byte. If no bytes available to read, Serial.peek () will return -1. This instruction has no parameters.

Serial.find () reads data from the input buffer (removing them from there) until the string (or a single character) be specified as a parameter, or have read all the data currently in the buffer. The statement returns "true" if the string is found or "false" if not.

4.8 Example: The following code uses Serial.find () to continuously read the data in the input buffer in search of the word "hello." You can test your typing behavior on the box of "Serial Monitor" chains we want to send to the plate. If the string "hello" is found, it is displayed by the "Serial monitor" the word "Found".

```
boolean encontrado;
void          setup(){
            Serial.be
            gin(9600)
            ;
}
void                          loop(){
            encontrado=Serial.find(
            "hola");  if  (encontrado
            == true){
                      Serial.println("Encontrado");
           }
```

Note: Here again we see an example of "if": basically check what is the value returned by Serial.find () is true in order to print "Found". It is important to realize that they have to write the two equal in the condition of "if" (we'll talk about it in the section).

Serial.findUntil () reads data from the input buffer (removing them from there) until the string (or a single character) be specified as the first parameter, or reach a final mark of search (which It is-or character-string specified as the second parameter). The statement returns "true" if it is the search string before the end mark search or "false" if not.

Serial.readBytes (): Read input buffer (removing them from there) the number of bytes specified as the second parameter (or if not enough bytes arrive, until it has passed the specified Serial.setTimeout () time). In any case, the read bytes are stored in an array -of "char []" - specified as the first parameter. This command returns the number of bytes read buffer (so a value
0 means that invalid data were found).

4.9 Example: The following code (where use is Serial.readBytes ()) will get what is in the input buffer 20 in 20 bytes, and stored in the array "myarray", then showing the amount of bytes and values obtained as a string. If more than 20 bytes in the buffer, the next repetition of "void loop ()" is rerun Serial.readBytes (), which reads as follows
20 bytes of buffer and the values that were in the array are overwritten by new. You can use the "Send" button "Serial Monitor" to send characters to the plate and observing the result.
ARDUINO. Training Workshop

```
char miarray[30]; byte bytesleidos; void setup(){
        Serial.begin(9600);
}
void        loop(){        bytesleidos=Serial.readBytes(miarray,20);
        Serial.println(bytesleidos);
        Serial.println(miarray);
}
```

Serial.readBytesUntil (): Read input buffer (removing them from there) the number of bytes specified as the third parameter, or if it is before a character string, or character individually specified as the first parameter that mark end, or if not get enough bytes or end mark is, until it has passed the specified Serial.setTimeout () time. In any case, the read bytes are stored in an array-of-type "char []" - specified as the second parameter. This command returns the number of bytes read from the buffer (so a value of 0 means that invalid data were found).

Serial.setTimeout () has a parameter (of type "long") it used to set the maximum number of milliseconds Serial.readBytesUntil instructions () and Serial.readBytes () will wait for the arrival of data to the serial input buffer . If either of these instructions it receives no data and this period is exceeded, the sketch continue execution on the next line. The default timeout is 1000 milliseconds. This instruction is usually written in "void setup ()". No return value.

Serial.parseFloat (): Read input buffer (removing them from there) all the data until it is a decimal number. Its return value - type "long" - decimal number that is then found. When it detects the first subsequent invalid character, stop reading (and therefore it will not eliminate the data buffer). This instruction has no parameters.

4.10 Example: The following code (with the help of the "Send" the "Serial Monitor" button) allows us to test the use of Serial.parseFloat ():

```
float numero;
   void            setup(){
               Serial.be
               gin(9600)
               ;
   }
void loop(){
   /* Vacía el buffer hasta
   reconocer
   algún número decimal o vaciarlo del
   todo */
               numero=Serial.parseFloat();
   /*Imprime el número decimal
   detectado,
   y si no se ha encontrado ninguno, imprime
               0.00 */ Serial.println(numero);
   /*Lee un byte más y lo imprime. Si se hubiera detectado un número
   decimal, ese byte sería el carácter que está justo después de él. Si el
   buffer está vacío porque Serial.parseFloat() no encontró ningún
   número decimal, entonces devuelve -1 */
               Serial.println(Serial.read());
   }
```

Serial.parseInt (): Read input buffer (removing them from there) all the data until it is an integer. -of Its return value type "long" - will be found then that integer. When it detects the first subsequent invalid character, stop reading (and therefore it will not eliminate the data buffer). This instruction has no parameters.

The objects range of other Arduino

So far we have assumed the use of Arduino UNO, which has only one TTL-UART chip and thus allows a single series object, called "Serial". However, other models have more Arduino objects of this type, and therefore a greater flexibility in the use of serial communication.

For example, Arduino Mega has four TTL-UART chips. This means we can use up to four series, called "Serial", "Serial1", "Serial2" and "Serial3" objects. The first remains associated with pins 0 and 1, the "Serial1" object is associated with the pin 18 (TX) and 19 (RX), the "Serial2" to pins 16 (TX) and

17 (RX) and the "Serial3" to pins 14 (TX) and 15 (RX). Each of these objects can be opened independently (writing Serial.begin (9600); Serial1.begin (9600); Serial2.begin (9600) or Serial3.begin (9600), respectively), and can also send and receive data independently. However, the only object also associated with the USB connection is "Serial" (because it is the only one connected to the converter chip ATmega16U2).

Arduino Leonardo also has, outside the "Serial" object, the object "Serial1". In this case it is possible to separate the two serial communication channels that can handle TTL-UART chip incorporated into the ATmega32U4: the "Serial1" object is allocated for the transmission of information through pins 0 (RX) and 1 (TX), and the "Serial" reserves subject to the same serial transmission but made through the USB-ACM communication (which in turn is different from the USB communication used for simulations of keyboard and mouse).

Keep in mind also that, unlike what happens with the model UNO, performing a skit on the Leonardo plaque is not reset when the "Serial Monitor" opens, so you can not see the data which they have been previously sent by the board to the computer (such as those sent in the "setup ()" with Serial.print () or similar function). To circumvent this problem, you can write the following line right after Serial.begin (): while (! Serial) {;} This will make while serial communication (ie, while the "Serial monitor is not open is not open "), the sketch do nothing and stay on hold" latent ". When we study the loop "while" at the end of this chapter will be better understood its meaning.

Arduino Due, in turn, has three additional serial ports (in addition to the port "Serial" located in the standard pins 0 and 1). These extra ports can be used in our sketches by objects "Serial1" (corresponding to pins 18 and 19 -tx- -RX-), "Serial2" (corresponding to pins 16 and 17 -tx- -RX-) and "Serial3" (corresponding to pins 14 and 15 -RX- -tx-). Of these, it is the "Serial" object which is connected to the USB mini B connector on the board (through the chip ATmega16U2), so that will be the object that we use in our sketches to send and receive data series through USB cable. We could send and receive serial data directly to SAM3X chip through the USB cable if we connect this to the mini-USB socket A, but in this case the object to use in our sketches should be another, called "SerialUSB".

In any case, keep in mind that the Arduino Due working all ports to 3.3 V (instead of 5 V as the rest of plates).

INSTRUCTIONS TIME MANAGEMENT

These instructions do not belong to any object, so it is written directly:
millis (): returns the number of milliseconds (ms) since the Arduino board began running the current sketch. This number is reset to zero after about 50 days (when the value exceeds the maximum allowed by its type, which is "unsigned long"). No parameters.

micros (): Returns the number of microseconds (ms) since the Arduino board began running the current sketch. This number-of type "unsigned long" - will be reset to zero after about 70 minutes. This statement has a resolution of 4 microseconds (ie, the return value is always a multiple of four). Remember that 1000 is one millisecond ms and therefore is a second 1000000 ms. No parameters.

delay (): Pause the sketch for the number of milliseconds specified as a parameter-of type "unsigned long" -. No return value.

delayMicroseconds (): Pause the sketch for the number of microseconds specified as a parameter-of type "unsigned long" -. Currently the maximum value that can be used with precision is 16383. For more expect this, we recommend using the instruction delay (). The minimum value that can be used with precision is 3 microseconds. No return value.

Example 4.11: A simple code of any of the above instructions is this.

```
unsigned long time;
void            setup(){
              Serial.be
              gin(9600)
              ;
}
void loop(){
            time        =
            micros();
            Serial.printl
            n(time);
            delay(1000);
}
```

If the above code is executed you can see the "Serial monitor" how will increase the time from which started the sketch. The observed value increases approximately one second each time.

Example 4.12: Another illustrative code is as follows.

```
unsigned long inicio, fin, transcurrido;
void        setup(){
        Serial.be
        gin(9600)
        ;
}
void                loop(){
        inicio=millis();
        delay(1000);
        fin=millis();
        transcurrido=fin
        -inicio;
        Serial.print(transcurrido);
        delay(500);
}
```

In the above code you can see a way to count the elapsed time between two points in time. The procedure is stored in a variable the value returned by millis () at baseline, and store in a different variable, the value returned by millis () at the final moment, and then subtract from each other and so determine the length of time elapsed (in the example should be approximately one second). This calculation in the above code is performed every half second.

INSTRUCTIONS MATHEMATICS, trigonometric, and pseudorandomness

The Arduino language has a set of mathematical instructions and pseudorandomness we can come in handy in our projects. These are:

abs () returns the absolute value of a number passed as parameter (which can be both whole and decimal). That is, if the number is positive (or 0) returns without altering its value; if negative, it returns "become positive. " For example, 3 is the absolute value of both 3 and -3.

min () returns the minimum of two numbers passed through parameters (which can be both integer and decimal).

max () returns the maximum of two numbers passed through parameters (which can be both integer and decimal).

Often used functions min () and max () to restrict the minimum or maximum value that can have a variable whose value comes from a sensor.
For example, if we have a variable called "sensVal" that derives its value from a sensor line sensVal = min (sensVal, 100); It ensures that this variable will never have a smaller value of 100, although the sensor receives in any given smaller values. In this sense, we may be useful another more specific instruction called constrain (), which serves to contain a specific value between two minimum and maximum extremes

constrain () recalculates the value passed as the first parameter (call it "x") depending on whether you are inside or outside the delimited by the values passed as the second and third parameters (call them "a" and "b" range respectively, where "a "always it has to be less than" b "). The three parameters can be both integer and decimal. In other words:

If "x" is between "a" and "b" constrain () returns "x" unchanged. If "x" is less than "to" constrain () returns "a"
If "x" is greater than "b" constrain () returns "b"

A more complex instruction that earlier (but most versatile) is the map () statement. This instruction is used in many projects to adapt the input signals obtained by different sensors optimal numerical range to work. We will see several examples of its use in later sections.

map (): change a -especificado as the first parameter-value which is within the range initially (delimited with -second minimum and maximum parameter-parameter--third) so that it is within a range (with another minimum - fourth parameter-parameter-Quintus and maximum) so that the conversion value is proportional as possible. This is what is called "map" a value: the minimum and maximum range change and so intermediate values are appropriate to that change. All parameters are of type "long", so negative integers, but no decimal numbers are also supported: if one appears in the internal calculations of the instruction, it will be truncated. The value returned by this instruction is precisely mapped value.

4.13 Example: The following code shows the use of map () statement.

```
void    setup(){
        Serial.be
        gin(9600)
        ;
```

```
        Serial.println(map(0,0,100,200,40
        0));
        Serial.println(map(25,0,100,200,4
        00));
        Serial.println(map(50,0,100,200,4
        00));
        Serial.println(map(75,0,100,200,4
        00));
        Serial.println(map(100,0,100,200,
        400));
        /*El valor puede estar fuera de los rangos,
        pero igualmente se mapea de la forma
        conveniente                              */
        Serial.println(map(500,0,100,200,400));
}
void loop(){}
```

If the output in the "Serial Monitor" is observed, you can see how the minimum value of the initial range (0) is mapped to the minimum value of the final ranking (200), the value 25 (a quarter of the initial span) is mapped to the value 250 (one quarter of the total end range), the value 50 (half the initial span) is mapped to the value 300 (half of the total end range), the value 75 (three quarters of initial span) is mapped to the value 350 (three quarters of the total end range), the maximum value of the initial range (100) is mapped to the maximum value of the final range (400) and a value outside the initial range (500) is mapped proportionally to other value out of the final rank (1200).

Observe in the latter case that the mapped values are not restricted within the new range because it can sometimes be helpful to work with outliers. If you want to limit them, use the constrain () statement or before or after map ().

See also the minimum of the two ranges (the current and modified) can actually have a value greater than the maximum. Thus, they can be reversed number ranges. For example, map (x, 1, 50, 50, 1); invest the original values: 1 would become 50, 2 ... 49 and 50 would become 1, 49 2 ...

For the curious: mathematically, the map () instruction performs this calculation: I xmapeado = (x- minactual) * (maxfinal - minfinal) / (maxactual - minactual) + minfinal where "x" is the value to be mapped, "minactual" is the minimum value of the current range, "maxactual" is the maximum value of the current range, "minfinal" is the minimum value of the final range and "maxfinal" is the maximum value of the final ranking.

The Arduino language also provides instructions regarding powers:
pow (): Returns the value resulting from raising the last number as the first parameter (the "base") to the last number as the second parameter (the "exponent" which may even be a fraction). For example, if we execute result = pow (2.5); the variable "result" worth 32 (25). Both parameters are of type "float" and the return value is of type "double".

If you want to calculate the square of a number (ie, the result of multiplying that number by itself, besides using the pow () statement citing the exponent number 2, you can use a specific instruction that sq () whose only parameter is the number (of any kind) to be squaring and the result is returned as a number of type "double".

sqrt () returns the square root of the number passed as a parameter (which can be both whole and decimal). The return value is of type "double".

We may also use trigonometric instructions. It goes beyond the scope of this book to explain the usefulness and mathematical meaning of these functions; if the reader wants to deepen their knowledge of trigonometry, I recommend consult the Wikipedia article entitled "trigonometric function".

sin () returns the sine of an angle specified as a parameter in radians. This parameter is of type "float". His return can be a value between -1 and 1 and is a "double".

cos () returns the cosine of an angle specified as a parameter in radians. This parameter is of type "float". His return can be a value between -1 and 1 and is a "double".

tan () returns the tangent of an angle specified as a parameter in radians. This parameter is of type "float". His return can be a value between $-\infty$ and ∞ and is a "double".

We may also use pseudorandom numbers in our Arduino sketches. A pseudorandom number is not strictly a random number according to rigorous mathematical definition, but for our purposes the level of randomness that reach the following functions will suffice:

randomSeed (): initialize the pseudorandom number generator. Is usually running on the "setup ()" section to use thereafter pseudorandom numbers in our sketch. This statement has a parameter of type "int" or "long" called "seed" that indicates the value at which begin the sequence of numbers. Equal seeds produce equal sequences so interested in multiple executions of randomSeed () using different seed values to increase randomness. We may also like sometimes the opposite: set the seed for the random number sequence is repeated exactly. It has no return value.

random (): After initializing the pseudorandom number generator with randomSeed (), this instruction returns a pseudorandom number of type "long" between a minimum value (specified as the first parameter -optional-) and a maximum value (specified as second parameter) minus one. If the first parameter is not specified, the default minimum value is 0. The type of both parameters can be any as long as whole.

Example 4.14: An example use of pseudo instructions would read sketch. There you can see that you have a seed fixed and pseudo-random numbers are generated between 1 and 9. Note that, having seed fixed, if the "Serial Monitor" opens at different times (assuming you use a plate UNO) in all the sequence numbers it will be exactly the same:

```
void    setup(){
        Serial.be
        gin(9600)
        ;
        randomS
        eed(100);
}
```

```
void                    loop(){
        Serial.println(random(
        1,10)); delay(1000);
}
```

Besides the above mathematical instructions, Arduino language has several arithmetic operators, some of which have already been appearing in some sample code. These operators work both as decimal numbers and integers are:

Arithmetic Operators

Operator + sum

- Subtraction operator

* Multiplication operator

/ Operator division

% Modulo operator

The operator module is used for the remainder of a division. For instance: 27% 5 = 2. It is the only operator that does not work with "floats".

A final comment: possibly a priori mathematical functions Arduino language may seem low compared with other programming languages. But nothing is further from reality: precisely because the language Arduino no longer a "makeup" of the C / C ++ language, we actually have at our disposal most mathematical functions (and indeed, virtually any type of function) that It provides the C / C ++ language. Specifically, we can use all the functions listed in the online reference "avr-libc" (http://www.nongnu.org/avr-libc/user-manual). So, if we need to calculate the exponential of a decimal number "x", we can use exp (x) ;. If we want to calculate the natural logarithm, we can use log (x) ;. If we calculate the logarithm base 10, you can use log10 (x) ;. If we calculate the module of a division of two decimals, we can use fmod (x, y); etc.

www.ingramcontent.com/pod-product-compliance
Lightning Source LLC
Chambersburg PA
CBHW080804180526
45168CB00006B/2327

* 9 7 8 1 5 1 7 6 4 9 8 6 9 *